Vita Mathematica
Volume 11

Edited
by Emil A. Fellmann

Evariste Galois
1811–1832

Laura Toti Rigatelli

Translated from the Italian
by John Denton

1996   Birkhäuser Verlag
       Basel · Boston · Berlin

Author's Address:

Dott. Laura Toti Rigatelli
Dipartimento di Matematica
Università di Siena
Via del Capitano, 15
I-53100 Siena

QA
29
.G25
T6713
1996

The biographical part is largely based on
Evariste Galois, matematica sulle barricate published in 1993
©RCS Sansoni Editore Sp.A., Firenze, 1993

A CIP catalogue record for this book is available from the Library of Congress, Washington D.C., USA

Deutsche Bibliothek Cataloging-in-Publication Data

Toti Rigatelli, Laura:
Evariste Galois : (1811–1832) / Laura Toti Rigatelli. Transl. by
John Denton. - Basel ; Boston ; Berlin : Birkhäuser, 1996
   (Vita mathematica ; Vol. 11)
   ISBN 3-7643-5410-0 (Basel ...)
   ISBN 0-8176-5410-0 (Boston)
NE: GT

This work is subject to copyright. All rights are reserved, whether the whole or part of the material is concerned, specifically the rights of translation, reprinting, re-use of illustrations, recitation, broadcasting, reproduction on microfilms or in other ways, and storage in data banks.
For any kind of use, permission of the copyright owner must be obtained.

© 1996 Birkhäuser Verlag, P.O.Box 133, CH-4010 Basel, Switzerland
for the English Version of the biography, the mathematical chapter and updated bibliography.
Printed on acid-free paper produced from chlorine-free pulp. TCF ∞
Printed in Germany
ISBN 3-7643-5410-0
ISBN 0-8176-5410-0

9 8 7 6 5 4 3 2 1

For Paolo

Contents

Preface	9
Preface to the English Translation	11
Introduction	13

1 The Early Years
    1811–1823 15
    A Strict School 20

2 The Excitement of Mathematics
    A Beneficial Injustice 29
    Early Dreams of Glory 33
    Two Serious Setbacks 42
    The Beginning of a Difficult Year 48

3 The "Three Glorious Days"
    The Ordinances of CHARLES X 53
    Riots in Paris 56
    Another King! 63
    The End of a School Career 68

4 To LOUIS-PHILIPPE!
    A New Plan 77
    Trouble with the Law 81
    The Trial of the Fifteen 95
    Prison 96

5 A Pointless Death   107

The Mathematical Work of EVARISTE GALOIS 115

Bibliography 139
Index of Names 153
List of Illustrations 161

Preface

Literary fiction, the cinema and theatre have all shown interest in EVARISTE GALOIS. He has been turned into a myth by mathematicians, though praise has not always been unanimous. However, neither uncritical biographers nor hostile critics have ever founded their arguments on reliable historical evidence. Since no thoroughly researched biography of GALOIS was available, I decided to try and fill the gap, starting off from documentation in archives, and evidence from contemporary memoirs and newspapers.

The period I spent in the splendid surroundings of the Historical Library of the City of Paris, housed in the Hôtel de Lamoignon, was an exciting one, full of new discoveries, I was able to share the same experience as that described by LEOPOLD INFELD in the introduction to his novel *Whom the Gods Love*: like him I fell in love with 19th century France.

On the basis of the analysis and interpretation of a series of hitherto neglected documents. I have been able to provide a new version of the circumstances leading to GALOIS' death. These documents are: the article that appeared in the Lyons newspaper *Le Précurseur*, and memoirs written by H. J. GISQUET, prefect of police, and LUCIEN DE LA HODDE, one of LOUIS-PHILIPPE's spies.

Over the last few years, I have tried to share my enthusiasm for GALOIS with many people with whom I have come into contact, above all my students, among whom I should like to single out: LAURA BETTI, BARBARA BONUCCI, CLAUDIA BORSELLI, FRANCESCO BUINI and BEATRICE TREMITI, my companions on a journey to Paris in search of the initial documentation of EVARISTE GALOIS. I am grateful for the long, stimulating conversations I have shared with them on this topic.

I have been provided with very helpful bibliographical references by IVOR GRATTAN-GUINNESS and ROBERTO G. SALVADORI, to whom I am very grateful. My thanks also go to RAFFAELLA FRANCI and PAOLO PAGLI for reading the first draft and making some useful suggestions.

Lastly, this book would never have been written without the crucial contribution of my daughter FIORENZA. Her name deserves to appear as co-author.

Siena, October 1992

Laura Toti Rigatelli

Preface
to the English Translation

This English translation of my biography of GALOIS, which appeared in Italian under the title: *Matematica sulle barricate* (Mathematics on the Barricades), has allowed me to make a few additions. I have added a section to chapter 4 on the trial of the fifteen, and a new final chapter on GALOIS' contribution to mathematics. The bibliography has been expanded to include works on GALOIS' "classic" theory of algebraic equations.

I should like to thank my son VIERI, ANDREA SORBI and PAOLO PAGLI for their help.

Siena, July 1995

Laura Toti Rigatelli

Introduction

The *Société des Amis du Peuple* gathered together its most active members in the republican party in the spring of 1832. Their sole enemy was the "King of the French," and their sole aim to put an end to his reign. In order to achieve this, they were ready for any kind of violence.

They had not only been disappointed by the revolution of July 1830 that had placed Louis-Philippe d'Orléans on the throne, but saw it as something of an insult. Republican voices had called the people of Paris to the barricades, and republican blood had been shed on them, but the end result was that, next to the ageing General Lafayette, yet another king had appeared on the balcony of the Hôtel de Ville to greet the crowds below. While the streets of Paris were still strewn with the bodies of those who had given their lives to rid France of Bourbon tyranny, the republicans had already begun preparations for a new uprising. They had been plotting, scheming and planning for nearly two years. Even sacrificing one of the great Parisian landmarks, the Cathedral of Notre Dame, was considered worthwhile. Setting fire to the twin towers of the façade was to be the signal for the call to arms. While flames were consuming the spires of the Cathedral, the conspirators would spread through the streets of Paris inciting the people to revolt. The more fanatical among them were convinced that blaming the fire on the royal police and the gathering together of a huge crowd would have been enough to spark off a new revolution.

However, Lucien De La Hodde, a spy who had infiltrated the Society, had warned Gisquet, the prefect of police, causing the ingenious plot to fail. Notre Dame was saved, and with it Louis-Philippe's throne.

Despite all this, in that spring of 1832, the group of revolutionaries saw the people of Paris, many of whom had been ruined by a serious economic slump, and also struck by a terrible cholera epidemic, as ready to respond to the call for a new uprising. On 7 May, the leading members of the Society met at 18 rue de l'Hôpital-Saint-

Louis. It was now time to make a decision and prepare a new, foolproof plan, which would turn out differently from the tragic failure of two years previously. One of those present was a twenty-year-old by the name of EVARISTE GALOIS, who had been sentenced to nine months for his political opinions. Prison records described him as a "mathematics tutor."

"Thoroughly disillusioned, even with love and fame," it was he who had an idea which led to a detailed plan. He often used to say: "if I were only sure that a body would be enough to incite the people to revolt, I would offer mine." About one month later, on 5 June, the streets of Paris were once again blocked by barricades. For the first time, flying from them was the red flag of socialism.

GALOIS, whose idea had triumphed, was in no position to fight his perpetual enemies. He had died on 31 May.

1 The Early Years

1811–1823

The GALOIS family had embraced revolutionary ideals from the very beginning of the revolution in 1789. They lived in Bourg-la-Reine, in the department of Hauts-de-Seine, 4 km south of the Porte d'Orléans on the outskirts of Paris. Their village had grown out of a subsidiary establishment of the Abbey of Montmartre, founded by LOUIS VI's queen, ADELAIDE OF SAVOY, in the 12th century.

Ownership, since the mid-18th century, of a school had contributed to both the family's prosperity and prestige. When the winds of revolution changed Bourg-la-Reine into Bourg-Egalité, the old church-run schools were abolished, but not initially replaced by sufficient secular schools. Consequently, the GALOIS family establishment began a new lease of life, that lasted until the end of the Napoleonic era.

The brothers THÉODORE-MICHEL and NICOLAS-GABRIEL GALOIS were both loyal subjects of the Emperor NAPOLEON. The former showed his devotion by joining the army and becoming an officer in the Imperial Guard, while the latter, more interested in books than battlefields, decided to take over management of the family school. The ease with which he wrote plays and verse made the clever, witty NICOLAS-GABRIEL GALOIS a great success in the *salons*, and his kind, courteous behaviour fascinated his circle of friends and acquaintances.

Almost opposite the GALOIS, in the Grand Rue, the main street of Bourg-la-Reine, which is now called Avenue du Général Leclerc, lived the DEMANTE family. THOMAS-FRANÇOIS DEMANTE was an *agrégé* in law from the Sorbonne and presiding judge at the law courts in Louviers, and, though considerably older than him, a friend of NICOLAS-GABRIEL GALOIS. All the latter had to do to find a wife was cross the road, having chosen ADÉLAIDE-MARIE, probably the prettiest of DEMANTE's daughters, as his future bride. When they married in 1808, NICOLAS-GABRIEL GALOIS was thirty and his bride only twenty.

ADÉLAIDE-MARIE was far from being an ordinary woman. She had received a rigorous education from her father, a first class Latin scholar. She was intelligent, lively and generous and had a strong character, her approach to life being greatly influenced by examples taken from her classical studies.

The marriage was a happy one. After a year, a daughter was born, who was called NATHALIE-THÉODORE, in honour of her uncle away fighting on military campaigns in Europe. Two years later, at 1 a.m. on 25 October 1811, the first son was born. When choosing a name, this time his parents consulted the calendar. The following morning, his proud father informed LAVISÉ, Mayor of Bourg-la-Reine, that his son's name for the registry was EVARISTE, after the saint whose feast day falls on 26 October.

The celebrations in the GALOIS family for the birth of EVARISTE, though as lavish as the family budget would allow and their social position required, were undoubtedly insignificant when compared with those in the month of March of that same year accompanying the birth of the son of NAPOLEON and MARIE-LOUISE OF AUSTRIA and heir to the Imperial throne. A 101 gun salute had greeted the birth of NAPOLÉON-FRANÇOIS-JOSEPH-CHARLES. His father had conferred the title of "King of Rome" on the child, to underline the universal character of his imperial claims.

It really did seem, in that happy year, that NAPOLEON's fortunes would never wane, and that the BONAPARTE dynasty would become the most powerful in Europe. The Emperor had not seen the end of the Franco-Russian alliance as a threat, but as a spur to further conquest. War between the two powers, in his view, would not only allow him to reach and conquer Moscow, but would also have opened up the way to India, which he considered the real source of Britain's wealth.

In the hope of new glory, then, on 24 June 1812, at the head of his Great Army, NAPOLEON crossed the River Niemen, the natural frontier between the empires of east and west. But only four months later the retreat began. After one or two more battles, which had given him a few temporary victories, the Emperor was finally defeated by the intervention of Austria, and forced to abdicate unconditionally on 6 April 1814. Under a month later, on 3 May, LOUIS XVIII, brother of the guillotined LOUIS XVI, entered Paris. The white Bourbon flag flew again from the towers of Notre Dame.

The Early Years

Fig. 1: Plaque on the house built on the site of GALOIS' birthplace

The *Marseillaise* was forbidden. It only took the French people a few weeks to realize that the King's promise of a constitutional monarchy was only an expedient to trap them into believing that the main achievements of the Revolution would not be lost.

LOUIS XVIII was firmly convinced that he was King by divine right, and managed to concentrate power in the hands of the monarchy, leaving the people with practically no political representation. The members of the Chamber of Peers were directly nominated by the King, and membership of the Chamber of Deputies was restricted to a specific social class, since an income on which 300 francs in direct taxation were due was required for the right to vote, and the figure rose to 1,000 francs for candidates for election. These regulations still left the upper middle classes, who were increasingly irritated by the exaggerated claims of the returned *émigré* aristocracy, dissatisfied, while the renewed influence of the clergy was a cause of unrest for all social classes. Just a few months of the new régime were enough to create such a climate of dissatisfaction that the people gradually became ready for another revolution. The restoration of the Bourbons provoked the rise of a new, nationwide liberal movement. In Bourg-la-Reine, NICOLAS-GABRIEL GALOIS was unanimously recognized as its local leader.

Meanwhile, NAPOLEON in exile on the island of Elba, was waiting for the right occasion to return to power. On 26 February 1815, accompanied by the notes of the *Marseillaise* and shouts of "Vive l'Empereur!", he set sail on the brig *l'Inconstant*. He landed in Provence on 1 March, and, twenty days later, acclaimed by cheering crowds, took up residence again in the Tuileries. One of the repercussions of NAPOLEON's return to Paris was the appointment of NICOLAS-GABRIEL GALOIS as Mayor of Bourg-la-Reine.

However, the dreams of NAPOLEON and his supporters only lasted one hundred days, ending abruptly on the battlefield of Waterloo, where the French army was defeated by combined Anglo-Prussian forces. On 8 July LOUIS XVIII returned to the Tuileries. The restored Bourbon régime now took a turn for the worse, as compared to the situation three months before. Nevertheless, in Bourg-la-Reine the situation remained stable, and EVARISTE's father was reconfirmed as Mayor. His predecessor, whom the final collapse of NAPOLEON's régime should have restored to office, had been involved in some shady business dealings, and had been forced to leave the small town. In any case, NICOLAS-GABRIEL GALOIS' great popularity was enough to defeat extreme monarchist opposition, and keep him in office.

It was not so much the King's own behaviour as that of his entourage that lay behind the new political climate throughout the country. LOUIS XVIII was not particularly interested in politics generally, and left affairs to his ministers. His greatest worry was his health. Fat and gouty, he could not even walk without help.

France was split between two warring factions. On the left were the most fanatical supporters of the ideals of the Revolution, and on the right, the royalists. Moderates were a minority. Their opponents called the royalists (or legitimists) "ultraroyalists" or "Ultras" for short. Their leader was the King's brother, the comte D'ARTOIS.

LOUIS XVIII had been fiercely criticized by the extreme right for having granted a constitution, or *Charte* (charter), as it was called, in order to evoke traditional connotations. The legitimists believed in a close alliance between the Catholic Church and the State, and were determined to punish all those "traitors" who had been involved in NAPOLEON's Hundred Days. Thus they began a series of punitive actions, that came to be known as the "White Terror", during which hundreds of Bonapartists were murdered. So as to be able to act undisturbed, they had no qualms about setting up secret societies, the most important of which was called the *Congrégation*. Many state officials owed their appointments to the sinister influence of this secret society. The legitimist societies also took on the task

Fig. 2: LOUIS XVIII

of religious propaganda, through fervent sermons in churches, and somewhat macabre ceremonies of reparation for the crimes of the Revolution.

It was only in 1818 that a left-wing party was actually set up. It took the name of *Indépendents* and included the whole of the opposition: ex soldiers of the Empire, liberals, republicans and Bonapartists. This opposition group did well at the 1819 elections, but their efforts turned out to be in vain, as a result of the reactionary counter measures following an event that had strong repercussions for the legitimists. While leaving the Opéra, in rue Richelieu, on the evening of 13 February 1820, the duc DE BERRY, son of the comte D'ARTOIS, and sole heir to the throne, was assassinated by a working man called LOUIS-PIERRE LOUVEL. The Ultras decided to revenge the death of the man in whom they had placed their hopes, putting pressure on the King to punish the whole opposition. The first repressive measure was a new electoral law restricting citizens' participation in public life even further. The richer electorate was given two votes. The spread of liberal ideas also had to be curtailed.

Though more than 50% of the population was illiterate, keeping up to date on events through conversations in taverns and markets, a number of anti-régime newspapers had been started up. The one considered the most dangerous was certainly *Le Censeur*, which, in its unashamedly liberal, secular approach, revived the anti-religious ideas of VOLTAIRE. The legitimists also asked the King to stamp down on the press, and he got a law through Parliament reintroducing press censorship. From then onwards the King's approval would be necessary for any publication.

The state schools were subject to strict control, and teachers suspected of anti-royalist leanings lost their jobs, their places being taken by priests. All this was happening at the time EVARISTE GALOIS was about to go to school in Paris.

A Strict School

ADÉLAIDE-MARIE GALOIS was a very well read lady, and it was this that allowed her to take charge of the education of her three children (in 1814 she had given birth to another son, ALFRED). Her mother's teaching was enough for NATHALIE-THÉODORE, the eldest of the three, before she got married, but her brothers required more careful attention.

After EVARISTE's tenth birthday, his parents decided the time had come to send him to school. They chose a college in Reims, to which, after passing the entrance examination, EVARISTE was admitted with a partial grant. However, when the time came for him to leave, his mother did not have the heart to let her gentle, shy little son go. He looked far too small and defenceless to be sent off so far from home. She decided to keep him at home and give him more time to play and enjoy their quiet family life. It was only two years later, in 1823, that EVARISTE's parents decided that the decision had to be taken, and he left his parents' home. He had been admitted to the fourth class as a boarder at the famous Paris *Lycée Louis-le-Grand*, despite his tender age.

Lycées had only been relatively recently set up in France. They were the most innovative item in the FOURCROY Law, which had been approved on 1 May 1802. The first year was called the sixth class and the last the first class, which was also called *rhétorique*, and was followed by a class called *philosophie*.

The *Louis-le-Grand* institute had been turned into a *lycée* at the beginning of the century, after a long and glorious history beginning

The Early Years

Fig. 3: The façade of the *Lycée Louis-le-Grand* before rebuilding in 1885

in 1563, with the transfer of the Jesuit school called *Collegium Societatis Jesu* to the Hôtel de Langres at 6 rue Saint-Jacques. By the end of the seventeenth century, the school had become so famous all over France that LOUIS XIV decided it was worthy to bear his name. It was the only educational institution in Paris to remain open during the Revolution. When all the other schools were in a state of chaos, the *Louis-le-Grand* had managed to maintain its reputation for academic excellence, especially in classical studies, despite all the changes in educational method and approach that were continually taking place. When, in the autumn of 1832, EVARISTE enrolled, the ancient Hôtel de Langres was in a very bad state of repair; large cracks in the walls contributed to its run-down appearance, and the gates and bars at the windows made it look more like a prison than a school.

There was a sharp contrast, however, between the gloomy appearance of the building and the liveliness of its young inmates, which had almost gone too far during NAPOLEON's Hundred Days. Many of the pupils would have liked to leave the school to fight with NAPOLEON. A riot broke out when they were forbidden to leave their classrooms. The school cook was nearly hanged, and the deputy headmaster risked being thrown out of a window.

There had been many episodes of rebellion and disobedience since 1815. The headmasters, LOUIS-GABRIEL TAILLEFER and his younger successor FRANÇOIS-CHRISTOPHE MALLEVAL, had been hard put to calm down their turbulent charges. Violent discussions and quarrels were the norm during recreation; hardly surprising, seeing that the school was attended by the sons of Jacobins, Bonapartists, legitimists, and supporters of practically all the other political factions. A further factor contributing to this strong inclination to rebellion was the strict discipline enforced in the school.

The pupils' day began early. At 5.30 a.m., a bell was rung in the unheated dormitories, each containing forty beds placed at exactly one metre from each other. After a quick wash at the courtyard fountain (the only one in the whole establishment), the pupils had to put on their uniforms in silence. The uniform, consisting of a dark blue military-style jacket and trousers, with a light blue collar and embroidery, and a two-cornered hat, had been chosen by NAPOLEON, who loved to look after even the tiniest details. The new régime of LOUIS XVIII had seen no reason to make any alterations.

Once dressed, and after assembly prayers, the pupils had to go straight to their classrooms. A period of study would increase their appetite for breakfast. The classrooms had no desks, but only steps, on which the pupils sat with their books and exercise books in their laps. Lighting was limited to one candle every two pupils. Rats swarmed over the floor and steps, occasionally even biting pupils. The wooden teacher's desk was on a high podium, rather like a pulpit, so that the teacher had complete control over class discipline and attention. In every classroom, a marble bust of either HOMER or CICERO looked down over the teacher's desk, on which a candle was placed. Classrooms, like the administrative offices, were heated by large stoves, which gave off more irritating smoke than heat.

At 7.30, after about two hours' study, breakfast was brought directly to the classrooms. It consisted of water and dry bread, and had to be eaten quickly and silently. The time limit was a quarter of an hour, and nobody was allowed to put pieces of bread in his pocket to eat later.

The above description should not lead readers to believe that this kind of breakfast was typical of the Paris of the time. Middle class Parisians had milky coffee, or, after the Restoration, when England became fashionable, tea. The rich might even have hot chocolate and freshly baked bread rolls with what BALZAC called *frippe*, i.e., butter, honey and jam. Why then were these growing boys fed on

The Early Years

a prison diet? It was probably believed that a strict diet would help form a strong character.

At 8.00, the dayboys arrived and lessons began, lasting until midday. Lunch was served in the refectory, and again eaten in absolute silence. During the meal, one of the tutors would read extracts from morally uplifting writings, on which the boys could later be questioned. This is how silence during the meal was obtained. Only three quarters of an hour were allowed for lunch, consisting of a gruel, to which fat had been added, meat and green vegetables. Fish occasionally took the place of meat, eggs being served with it in place of the greens. On Sundays, a special treat, in the form of a rice pudding, was the reward for the most diligent pupils. At last the recreation period arrived, during which the *lycéens* were allowed to talk and walk, not run, which was considered undignified for the over fifteens, around the courtyards.

The afternoon lessons began at 2.00 p.m. and lasted until 6.00, with a short break for a snack at 4.30. The dayboys went home at 6.00, but the boarders had another task to perform. Again in silence, they had to go to a service in the chapel, at which all their movements, including genuflecting and crossing themselves with holy water, had to be performed with military precision. At 7.30, after the service, dinner was served in the refectory, and bedtime, with no further time allowed for recreation, was at 8.30. The pupils also had to get ready for bed in silence.

It is hard to believe that, in a school with about 500 pupils, silence was really as constant as regulations required. Breaking the rules, which was, understandably, very frequent, was severely punished, and consisted of being locked in a punishment cell and fed only bread and water. In 1824, the *Lycée Louis-le-Grand* was the Parisian school with the largest number of these cells: twelve, to be precise. They were tiny, badly lit, damp rooms whose only furnishings were a wooden bench and a jug. The "prisoner" was shut in at 10.30 a.m. and left until 8.00 p.m., without even a candle in winter. Punishment usually lasted for a minimum of four days, and the pupil was given a long translation to do from Greek or Latin.

Pupils could be punished even for the smallest infringements. Apart from talking during compulsory silence, it was enough to take too large a helping at meals, or refuse unwanted food, break a plate or other crockery, turn over too much in bed, fidget in class or chapel, or get up to the usual schoolboy pranks. There were never vacancies in the punishment cells.

Fig. 4: Pupils' uniforms at the *Lycée Louis-le-Grand* 1806–1906

When EVARISTE arrived, MALLEVAL's place as headmaster had just been taken over by NICOLAS BERTHOT, a mathematics teacher, who had taught descriptive geometry at the *Ecole Polytechnique*, but whose notoriety is principally based on his treatment of the pupils at the *Lycée*. As soon as he arrived, rumors began to circulate that he had been put there by the legitimist *Congrégation*, with the purpose of preparing the way for a Jesuit takeover of the school.

The majority of the pupils were of the opinion that some demonstration was needed against this manœuvre. They decided to show their opposition by refusing to sing during the chapel services. Punishment was much more severe than the usual time in a cell. The ringleaders were identified and expelled. The result was a decision in favour of a more drastic act of rebellion, when the opportunity arrived. This came with the annual celebration, on 28 January, of the feast of the Blessed Emperor CHARLEMAGNE, the great reformer of education in 9th century Europe. It was the custom on this day, in all French schools, to hold a celebratory banquet, to which all the teaching staff and the best pupils were invited. Latin and French poetry was read, official speeches made and toasts proposed.

The Early Years

On Wednesday, 28 January 1824, seventy five of the best pupils from the whole school were invited, though not including EVARISTE, who had hardly had the time to make a name for himself. The room where the banquet was held was brightly lit (for a change), and decorated with white flags bearing the Bourbon fleur-de-lys. As soon as the pupils came in, BERTHOT realized that something was wrong. There was dead silence, something that it was usually impossible to obtain, even during school hours, on an occasion when the opposite was to be expected and was actually allowed. Nevertheless, he pretended not to notice, and the banquet got under way, until the moment came for the customary loyal toast. At this point, the pupils did not respond, only covering the headmaster's words with laughter. The immediate expulsion of the seventy five disloyal subjects deprived the *Lycée* of its most promising pupils.

Thanks to the excellent preparation he had received from his mother, EVARISTE, although not taking to the harsh discipline at the *Louis-le-Grand*, soon made his mark among the other pupils. He was awarded a prize and three distinctions at the end of the fourth class. The following year, he took part in the *Concours général*, an annual competition between the best pupils in Paris, and received a distinction for his translation from Greek, while at the *Lycée* he had won the first prize for Latin poetry. His mother had good reason to be proud of her son.

During the 1825–26 school year, EVARISTE was troubled by serious earache, lasting throughout the winter, which was no doubt caused by the cold, damp rooms of the Hôtel de Langres. Nevertheless, the four distinctions he was awarded at the end of the second class were a sure sign of excellent progress. Being one of the best pupils, however, was not enough to make EVARISTE happy. He greatly missed his cheerful father, to whom he was very attached, and the inevitable days spent in the punishment cell depressed and humiliated him in his solitary confinement. If the monotony of the routine at school was a source of misery for EVARISTE and his fellow schoolboys, events in the world outside were hardly designed to cheer up the rest of the French people.

On 16 September 1824, after several days of suffering, LOUIS XVIII died. He was the last King of France to be buried in the Abbey church of St. Denis. His successor, who took the name of CHARLES X, was his younger brother and leader of the legitimists, the comte D'ARTOIS. The new King's coronation in Reims Cathedral the following May confirmed the submission of the Crown to the

Fig. 5: CHARLES X and his generals

Catholic Church. The ritual of the *ancien régime* had been followed even down to the minutest detail.

So as to placate his subjects, at least temporarily, CHARLES had been forced to accept the *Charte* signed by his predecessor, and had abolished press censorship as a conciliatory gesture to left-wing groups. However, he had been just as willing to accept the ever more pressing demands of the clergy and Ultras. A significant concession to the Church was the introduction of the concept of crime against religion into the penal code. From now on sacrilege in churches would carry the death penalty. Although this law was of very limited applicability, proof being difficult to obtain, it did represent a serious statement of principle against the secular character of the State. Opposition now concentrated on religious affairs and the intrigues of the Jesuits, who were rapidly gaining control of education, consequently becoming the objects of ferocious satire and lewd songs.

The accession of CHARLES X also brought a small change to the *Lycée Louis-le-Grand*: a new uniform. The *lycéens* had been proud of their old military-style uniform. Now they felt rather ridiculous in their new round shaped hats and long trousers, which had replaced the *culottes* they were used to.

2 The Excitement of Mathematics

A Beneficial Injustice

During the summer of 1826, yet another headmaster arrived at the *Lycée Louis-le-Grand*. BERTHOT's position was given to PIERRE-LAURENT LABORIE, an almost sixty year old protégé of the *Congrégation*, who had previously taught theology at the University of Perpignan. A few days after taking up his new post, LABORIE sent a report to the Minister of Education, to explain how difficult a task he thought he would have:

> The pupils have little interest in religion. The minority whose faith is still intact are ashamed of crossing themselves, for fear of the sarcasm and laughter of their school fellows. Nothing is sacred for them. They have the spirits and hearts of savages. Their irreverence has reached its climax, and there is little hope that things may improve. Even the teachers set a bad example, by not attending chapel regularly. Parents set a bad example, by fuelling their sons' imagination and filling them with the spirit of rebellion, with their attacks on the Jesuits and talk about the danger of domination by the Church. The Jesuits are a favourite subject of conversation among our pupils. How can one cope with young people who are sure that their rebellious acts will meet their parents' approval?

The new headmaster was not a man of great intellect, and owed his career to political connections. He soon provided evidence of his narrow view of educational matters.

That autumn of 1826, at the age of only fifteen, GALOIS should have started the *rhétorique* class. LABORIE had no doubts that he was far too young for such a demanding experience, and would have to repeat the second class. His excellent results were irrelevant; everything had to be done at the appropriate time! This was the decision of which the stubborn headmaster informed EVARISTE's parents, in a letter, dated 21 August, addressed to his father. He underlined the fact that: "intelligence and spirit can compensate for study, but they cannot replace judgment that only comes with maturity." Foreseeing that EVARISTE would protest, he added: "He should be on his guard, since his new rivals will not make life easy for him. He will have to cope with one of the best classes in this school, and he will have to study very hard indeed, if he wants to remain one of the best pupils."

EVARISTE's father was extremely irritated by this letter, and strongly opposed LABORIE's foolish decision. Thus, in that autumn, as planned, GALOIS enrolled in the *rhétorique* class. At the end of the first term, his teacher, DESFORGES, wrote that he was studying *with zeal* and his conduct was *good*. However, probably under pressure from LABORIE, the following note also appeared in the report: "His mind is still too immature to be able to take full advantage of the *rhétorique* class."

In January, following further pressure from the headmaster, the GALOIS family had to give in. GALOIS returned to the second class taught by SAINT-MARC-GIRARDIN, who had been appointed to the *Louis-le-Grand* the previous September. It was all very humiliating for him, though, it must be admitted, it was the very fact of a stubborn man having his way that encouraged a new inclination in this promising young man's mind, to the extent of providing material for a real legend created by some of GALOIS' biographers.

The French school curriculum established in 1814, and partially modified in 1830, introduced compulsory mathematics into the third class, continuing in the *philosophie* class. In the first year, pupils studied notions of arithmetic, and the teacher could choose from several textbooks: *Arithmétique* by BÉZOUT, or the books by abbé BOSSUT, abbé MARIE, or LACROIX.

LACROIX's book was also prominent in the second class, where arithmetic was replaced by algebra and geometry. As an alternative to LACROIX, for the latter subject, the teacher could also opt for a book that had appeared for the first time in 1794, the year of the Terror, and which, in translation, was used throughout the 19th century by students all over Europe. We refer, of course, to the celebrated *Eléments de géometrie* by ADRIEN-MARIE LEGENDRE.

Apart from the importance of his research in mathematics, LEGENDRE, who was almost eighty by the time GALOIS had enrolled in the *Lycée*, was also well-known in France for his enthusiasm for the ideals of the Revolution. Though he had not taken a direct part in political events, he had approved of all the mathematical projects worked out during the Revolution. He had been a member of the Committee for Weights and Measures, set up in 1790 by the

Académie des Sciences, at the request of the Constituent Assembly. LEGENDRE had measured the length of the earth's meridian, with an accuracy that did not fail to astonish his colleagues on the committee, thus allowing the calculation of one metre as a ten millionth of the distance between pole and equator. His *Eléments de géometrie* had originated in the search for scientific rigour that had characterized French mathematics during the last decade of the 18th century, and which had encouraged LEGENDRE to illustrate geometry "more or less in the order of EUCLID." This text made its author internationally famous.

During the *rhétorique* class, pupils continued to study algebra and geometry with the same textbooks as the previous year. Five hours per week were devoted to mathematics in the third and second classes, going down to four in the *rhétorique* class. The courses in the first three years were called *Mathématiques préparatoires* or *Mathématiques élémentaires*. The course held during the final *philosophie* class was called *Mathématiques spéciales*, which prepared pupils for entrance examinations to specific higher education institutions. During the *philosophie* class sixteen hours per week were devoted to scientific subjects, as against only seven for an arts subject like philosophy.

In the course called *Mathématiques spéciales*, pupils continued with algebra, using the classic treatise by EULER, to which the *Traité de la Résolution des Equations numériques de tous les degrés* by LAGRANGE was added. The so-called applications of algebra to geometry (i.e., modern analytical geometry) were also tackled, using the text by LACROIX followed by that by POULLET DE LISLE or the *Traité analytique des courbes et des surfaces du second degré* by BIOT.

In 1819, school inspectors AMPÈRE and RENDU suggested the *Traité géometrique des sections coniques* by ROBERTSON as an alternative, adding *Logarithmes* by LALAND and *Fragments sur l'Algèbre et la Trigonometrie* by REYNAUD.

GALOIS' first mathematics teacher was CHARLES-LOUIS-CONSTANT CAMUS, appointed to the *Louis-le-Grand* in November 1824. He does not appear to have been particularly interested in the subject during his first two years at school. It was a combination of his being made to return to the second class and the new textbook that led to his "discovery" of the subject. A new teacher of *Mathématiques préparatoires* in the second class, JEAN-HIPPOLYTE VÉRON, called VERNIER, the previous October, had chosen LEGENDRE's book in place of that by LACROIX. (After 1838 a book written by VERNIER himself was to be used.)

The legend we have already referred to tells that GALOIS read LEGENDRE's book, which was intended for a two year course, in only two days, and that it was like an adventure tale for him. He forgot his squalid surroundings, and suddenly discovered a wonderful world where everything is harmony and simplicity at the same time, a happy world in which to seek refuge.

From that moment on he only showed interest in mathematics. This meant that he began to neglect the other school subjects and isolate himself from his classmates. His behaviour at school changed, noticeably.

In the second term his school report complained of irregular conduct and spasmodic study. Here is the general opinion of him that ends the report:

> This pupil, with the exception of the last fortnight, during which he did study a little, has shown no interest in his subjects, except for fear of punishment. He has gone ahead from one punishment to another. On occasion, he left out the end of his exercises, other times he bungled them, and in the case of Latin compositions, limited himself to writing out the subject. His ambition, his frequently ostentatious originality, his bizarre character keep him isolated from his school fellows.

VERNIER's opinion was different: *assiduous, fruitful application*. VERNIER was a young man with little experience, and was also not a particularly gifted teacher. He had little imagination, monotonously repeating theorems and problems just as they were presented in LEGENDRE's text, without adding any further comment or fresh examples. These lessons were obviously not enough for GALOIS, who was often capable of finding different demonstrations to those suggested by LEGENDRE by himself, and whose mind was teaming with questions, which it would have been a waste of time putting to VERNIER. He soon began to realize that his capabilities separated him from his classmates, and began to read other books, dissatisfied with this one algebra textbook.

When, in the autumn of 1827, he was allowed to join the *rhétorique* class taught by PIERROT and DESFORGES, GALOIS had lost all interest in arts subjects as the opinions of his teachers clearly show:

> There is nothing in his work except strange fantasies and negligence;
> He always does what he should not do. He gets worse every day;
> Absent minded and a chatterbox;
> Bad conduct, aspires to be original.

VERNIER, who actually did not entirely understand GALOIS' new turn of mind, gave a positive opinion on the report at the end of the first term, writing: *excellent application and progress.*

From that moment on, GALOIS spent days pregnant with creative tension, and probably many sleepless nights, in complete isolation, accompanied only by the sarcastic laughter of his school fellows. His mind swarmed with formulas and theorems, leaving him either exhausted, happy or disappointed. He had worked on an important problem for two months, which had remained unsolved for almost three centuries, the discovery of a formula for solving fifth degree equations. He thought he had found this formula, only to realize that there was an error in his reasoning. However, this disappointment only made him concentrate even more and delve deeper into this wonderful new world. In the second term report, DESFORGES wrote:

> He is under the spell of the excitement of mathematics. Therefore, in my opinion, his parents would be well advised to let him devote all his energies to this one subject. He is wasting his time here, and does nothing but irritate his teachers and force them to punish him continually.

Though a rather poor teacher, and perhaps for this very reason, VERNIER was aware that GALOIS, overwhelmed by his enthusiasm and incredible imagination, was not studying methodically, or at least without doing the necessary routine exercises and learning the standard rules that only too often mathematics teachers make their pupils repeat like a kind of catechism. But GALOIS had a low opinion of VERNIER and took no notice of his advice. He was now set on gaining admission to the *Ecole Polytechnique*.

Early Dreams of Glory

All academies and universities had been abolished during the Jacobin Terror, for fear of replacing the old hereditary aristocracy with a new intellectual one. The liberal-bourgeois turn following the events of *thermidor* 1794, on the other hand, encouraged new legislation that was to have important repercussions on French education: the creation of the *Ecole Normale* and the *Ecole Polytechnique*. The latter, which soon became one of the most prestigious scientific educational institutions in France, perhaps in the world, was, and still is, a preparatory institution for specialization in civil and military engineering.

In 1794 a Committee of Public Works was set up with the purpose of founding an institute for the training of engineers. One of its members was the mathematician GASPARD MONGE, 48 at the time, and at the peak of his scientific and political career. He had been a cadet at the military academy in Méziers and an influential Jacobin. Subsequently, he became a loyal Bonapartist and accompanied NAPOLEON to Egypt. Like LEGENDRE, he had been a member of the Committee for Weights and Measures, which concluded its mandate in 1799. As Minister for the Navy, he had been among those who signed the official documents concerning the trial and execution of LOUIS XIV.

The original, significant name of the new institution, which was planned along military lines and soon in operation, was *Ecole Centrale des Travaux Publiques*. The following year this was changed to its present name of *Ecole Polytechnique*. MONGE, one of its enthusiastic founders, was also its first director, and one of its most renowned teachers, together with other well-known scientists, such as LAGRANGE, LAPLACE, LEGENDRE, LACROIX and DE PRONY. The *Collège de Navarre*, in the present day rue Descartes, was chosen to house the school.

The *Ecole Polytechnique* introduced profound changes into the French educational world. Interest in pure research was successfully combined with the technical sphere, which was shown to require a solid theoretical foundation. A large group of promising youngsters, selected on the basis of a stiff entrance examination, and subjected to strict discipline, thus had the chance of studying the most advanced scientific disciplines. Teaching in the *Ecole* was centred on mathematics. Besides, in GALOIS' time, it was attended by young men with mostly liberal ideas, who were said to shake both Cross and Church with their songs, their motto being "For my country, for science and glory."

The *Ecole Polytechnique* was the only institution GALOIS really wanted to attend, and its discipline the only one he would willingly accept. All he had were his reading and imagination, and, without even telling his parents, he took the entrance examination in June 1828, but failed. Disappointed but not beaten, he decided to try again the following year, and, in the meantime, returned to the *Lycée Louis-le-Grand*, to attend RICHARD's course on *Mathématiques spéciales*. His meeting with LOUIS-PAUL-EMILE RICHARD, who immediately realized that he had a very gifted new pupil, was one of the happiest in GALOIS' life. At last, he had found someone capable of understanding his enthusiasm and ambitions.

The Excitement of Mathematics

RICHARD was born on 13 March 1785 in Rennes. His father was a lieutenant-colonel in the artillery, and had served with distinction in the armies of both Republic and Empire. A small accident in childhood having disqualified him from enlistment in the army, he had been unable to follow in his father's footsteps. Thus, to the great advantage of science, he embarked on a career in teaching. He began as a *maître d'étude* in the Imperial *Lycée* in Douai, moving on, the following year, to the Royal College in Pontivy, where he began to teach *Mathématiques spéciales* in 1816. In 1820 he moved to Paris, as a teacher of *Mathématiques élémentaires*, first at the *Collège St. Louis*, and then at the *Louis-le-Grand*. He was finally appointed to teach *Mathématiques spéciales* in the same school. RICHARD was not a creative thinker, but had great love for his subject which his considerable teaching gifts enabled him to communicate to his pupils. Many young students at the *Polytechnique* owed their initial interest in scientific research to RICHARD's lessons. They were captivated not only by their scientific content, but also by the refined mathematical language used by their teacher. RICHARD always kept abreast with developments in research, by reading the latest scientific articles and papers. Being very much in favour of the methods of projective geometry, when free from his teaching duties at the *Louis-le-Grand*, he would listen to the lectures given by CHASLES at the Sorbonne. He was a retiring, shy, very generous man, who was always ready to go out of his way to help his pupils. Apart from GALOIS, many other famous scientists had, not surprisingly, been his pupils. One of them was URBAIN LE VERRIER, the astronomer who used calculus to discover the position and existence of the planet Neptune, on the basis of disturbances in the movement of Uranus. Among others were the mathematicians CHARLES HERMITE and JOSEPH-ALFRED SERRET, the latter being the author of the first treatise on algebra to incorporate GALOIS' fundamental ideas.

In January 1837, RICHARD was awarded a medal for his services to state education. However, his written work, which is all didactically oriented, was never published and remained in the family. Nothing survives today. When he died, at the age of fifty-four, he had made such a name for himself that, although he had only been a school teacher, he was honoured by an obituary in *Nouvelles Annales de Mathématiques*.

GALOIS did not neglect his studies in his new class, though, understandably, he devoted most of his time to following up his intu-

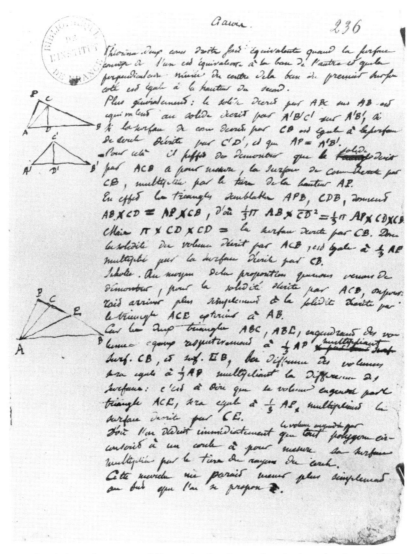

Fig. 6: Page from one of GALOIS' school exercises in the school year 1828–29. Fol. 236a of the manuscript

itions, with RICHARD's encouragement. He did his homework regularly and accurately, supplying such original solutions that RICHARD would base some of his lessons on them, without, however, suggesting they be imitated, such was the gap between GALOIS and the rest

The Excitement of Mathematics

of the class. His teacher kept these exercises throughout his life, and before dying, left this precious heirloom to CHARLES HERMITE, who was the only other person capable of appreciating their originality. They are now deposited in the library of the *Institut de France*.

RICHARD expressed himself very enthusiastically on GALOIS' work in all three terms of the school year, writing, among other things: "This pupil is far superior to all the others in his class," and: "This pupil only studies higher mathematics." Unfortunately not everybody shared RICHARD's enthusiasm. THILLAYE, the physics and chemistry teacher, had a poor opinion of GALOIS, which was certainly reciprocal.

JEAN-BAPTISTE-ANTOINE THILLAYE, a physician, physicist and chemist, curator of the collections and director of a physics laboratory at the Faculty of Medicine, was well-known in Paris, and his lessons, which were the end product of twenty years' experience, must have been high powered. The physics course covered statics, based on the treatises by MONGE and POINSOT, and astronomy, based on the *Abrégé* by DELAMBRE and the *Traité élémentaire d'Astronomie physique* by BIOT.

THILLAYE put himself out, just as much as RICHARD, for his pupils, even allowing them to use his physics laboratory at the University. However, GALOIS was only interested in mathematics, refusing categorically to show real interest in other subjects. This was seen as an eccentricity that caused several quarrels with the school authorities. THILLAYE's opinion can be summarized in his comment *"never studies"* on GALOIS' work over the three terms.

When GALOIS was engaged in lengthy, lively discussion with RICHARD, however, every trace of this alleged eccentric behaviour vanished. The two got on perfectly. The discussions between the enthusiastic teacher and his brilliant pupil covered the main concerns of mathematical research of the time, the names LEGENDRE, GAUSS, LAGRANGE and CAUCHY being the most frequently mentioned.

It was through the good offices of RICHARD that JOSEPH DIEZ GERGONNE accepted the results of GALOIS' first original research for publication in *Annales de Mathématiques*, the review he had founded in 1818, and which he still edited at the time. The article, which dealt with the demonstration of a theorem on recurring decimals, appeared in the 1 April 1829 issue. The author was described as "a pupil of the *Louis-le-Grand* College." It must have been highly unusual to read such a description of the academic status of the author of a scientific article. The article shows LAGRANGE's influence, and, although of high quality, gives no real idea of the future development of GALOIS' exceptional intuition.

On 6 April 1829, in the Norwegian town of Froland, the mathematician NIELS HENRIK ABEL died of tuberculosis, at the age of only twenty seven, and in a state of abject poverty. He had been born into a very poor, but well educated family, and had tried to earn a living by exploiting his exceptional talent for mathematics, but never succeeded. Just two days after his death, which originated in his miserable existence, a letter had arrived from Berlin appointing him professor at the University.

With the help of a small grant from the Norwegian government, ABEL had lived in Paris from July 1826 to March 1827, with the intention of meeting the greatest French mathematicians of the time. Unfortunately, he was very disappointed, because they did not pay much attention to him. The most famous among them was AUGUSTIN-LOUIS CAUCHY, who was about forty at the time, and at the peak of his academic and scientific career. He was a devout catholic and legitimist, who was perfectly satisfied with the political situation of the time. He was also an active member of the *Congrégation*. He taught mathematical analysis at the *Ecole Polytechnique* and was also a professor of mechanics at the Sorbonne and a fellow of the *Académie des Sciences*. He was not particularly generous to young scholars, and did not enjoy their company, preferring to devote all his time to his own research. In a letter sent to his teacher and friend BERNDT MICHAEL HOLMBOE, dated 24 October 1826, ABEL did not show much appreciation of CAUCHY:

> Cauchy is mad, and it is impossible to communicate with him. However, at the moment, he is the mathematician who knows how mathematics should be approached. His work is excellent, but difficult to read. At first I understood practically nothing, but I am getting somewhere now. Cauchy is a catholic bigot. That's very surprising in a mathematician! He is the only one dealing with pure mathematics here. Poisson, Fourier, Ampère etc. are only interested in magnetism and other areas of physics.

In the same letter he continued:

> I have completed an important *mémoire* on a certain class of transcendental functions for the *Institut de France*. The session will be on Monday. I showed it to Cauchy, but he just had a glance at it. Without being presumptuous, I think I can say it is good. I am curious to see what the opinion of the *Institut* will be.

The unlucky young Norwegian mathematician died before knowing the opinion he was so anxious to have. LEGENDRE and CAUCHY

Fig. 7: AUGUSTIN-LOUIS CAUCHY (1789–1857). An extremely productive mathematician, who made important contributions to infinitesimal analysis, by defining concepts allowing it to be treated with scientific rigour.

had been asked to examine the *mémoire*. LEGENDRE used the excuse that the handwriting was difficult to read, and left the task to CAUCHY, who, not being at all interested in the work of an unknown young foreigner, took it home, put it away somewhere, and forgot all about it.

In 1828, ABEL had written a paper in which he dealt, from a general viewpoint, with the problem of solving equations. This was a subject of considerable interest in algebra at the time, being a problem that had been keeping mathematicians busy for more than two hundred years. It was only in the 16th century that the formulas for solving third and fourth degree equations had been found. These particularly simple formulas had been obtained by only carrying out the four operations of arithmetic and those of root extraction, a finite number of times, on equation coefficients. Encouraged by these results, scholars studying algebra had set about searching for a similar formula for fifth degree equations, thinking that this would not be too difficult. After two centuries, during which all attempts had failed, in 1799, the Italian mathematician PAOLO RUFFINI demonstrated a surprising thesis: such a formula, i.e., "by radicals," in mathematical terms, did not generally exist if the degree of the equation was greater than four. ABEL, in 1824, had provided a different demonstration of RUFFINI's theorem. There were still many open problems, however, given that, in particular cases, the formula can be found, and ABEL presented two key points in the research proposal in his 1828 paper:
1. find all the equations of a specific degree that are solvable by algebra (i.e., by radicals);
2. decide whether a given equation can be solved by algebra or not.

Not only did time run out on ABEL in his ambitious project, but he did not even manage to publish the paper written in 1828. It was only printed ten years after his death. Thus, when GALOIS began to research the same problems, he could not have been aware of ABEL's work.

The spring of 1829 was undoubtedly the happiest period in GALOIS' life. The ideas flooding his mind were becoming clearer every day, and he began to see the solution of algebraic equations through the elaboration of new concepts, which he proudly realized were of enormous significance for future research. Not only was his research original, but it was destined to cause a revolution in algebra, an entirely new way of approaching the subject. He was well aware of this, and his enthusiasm was perhaps only less intense than that of RICHARD, who was now saying that the young GALOIS deserved to be admitted to the *Ecole Polytechnique*, without even having to take the entrance examination. He ought to be acclaimed as a genius, whose abilities where greater than those of his future teachers there. GALOIS was only seventeen at the time. The results of this highly productive period were two *mémoires*, which RICHARD believed could only be judged by the *Académie des Sciences*, the most prestigious scientific body in France.

The simplest way of presenting a *mémoire* to the Academy was to send the manuscript to the secretary's office, where it would be registered in the book of *entrées*, and, following that, included among the *mises à l'étude*. Only in exceptional cases was a different procedure followed. This consisted in finding a way of showing the work to a fellow of the Academy privately, who could vouch for at least the scientific worth of the subject, and present it to his colleagues during a session.

RICHARD had no doubts that this was the way to go about it. Momentarily forgetting his shyness, he approached the mathematician he believed to be the strictest, but most competent judge: CAUCHY. RICHARD had to put together all his courage, that was reinforced by his affection for GALOIS, and the conviction that his pupil had not only found the solution to the most important problem in algebra, but had also invented new, revolutionary methods of investigation in the discipline.

CAUCHY was everything but a friendly character, and showed little interest in other people's work. Since his nomination as a fellow of the Academy in 1816, he had only presented the results of his own research, with only one exception. This explains the astonishment of the fellows of the Academy, when CAUCHY presented GALOIS' *mémoire* entitled "Recherches algébriques," on 25 May 1829. FOURIER, NAVIER and CAUCHY himself were entrusted with the task of examining the *mémoire*, and passing judgment on it. One week later, during the session of 1 June, CAUCHY briefly presented GALOIS' second manuscript entitled "Recherches sur les équations algébriques de degré premier." This time judgment was entrusted to SIMÉON-DENIS POISSON as well as the presenter. The fellows nominated by the Academy, trusting their illustrious colleague, allowed CAUCHY to take the *mémoires* home with him.

Meanwhile, GALOIS was finishing his school year. RICHARD, naturally, gave him the first prize as the best pupil in his course, and had him take part in the *Concours général*. GALOIS' unorthodox approach to the problem to be solved for the competition, from a generalized viewpoint, was not what one would have expected from a schoolboy. For this reason he did not win the prize, but only came fourth. The first prize was awarded to AUGUSTE BRAVAIS, one of

GALOIS' classmates, who was later to apply the latter's fundamental mathematical ideas to the study of crystallography.

Two Serious Setbacks

As a follow up to the 1827 elections, which caused the fall from power of VILLÈLE, President of the Council of Ministers, JEAN-BAPTISTE MARTIGNAC, the Minister of the Interior became the King's spokesman before the two Chambers. MARTIGNAC ably managed to obtain majority support in his attempt to curb the demands of the clergy and issue ordinances against them. One of these ordinances prevented unauthorized members of religious orders from teaching, and this included the Jesuits.

CHARLES X, who favoured the restoration of the power of the Church, did not hide his disapproval of these measures and personal hostility to MARTIGNAC. Sure of the King's benevolence, the legitimists became overtly bold and insolent. This was also the situation in Bourg-la-Reine, where, at the beginning of 1829, a young priest was appointed to the parish, and soon formed an alliance with the local Ultras, particularly with one of the members of the local administration, who was unhappy about the presence of a liberal like NICOLAS-GABRIEL GALOIS as Mayor. The two decided to force him out of office, by circulating forged, vulgar epigrams, allegedly written by him.

A scandal broke out and EVARISTE's father was forced to leave Bourg-la-Reine and move to Paris, where he set up house in rue Jean-de-Beauvais. On 2 July, being unable to bear the situation any longer, when at home alone, he committed suicide.

His funeral was held in the church of St. Etienne-du-Mont in Paris, where, despite the fact that he had taken his own life, the priests agreed to receive the body. From there the funeral cortège moved on to Bourg-la-Reine. Despite the hostility of a group of Ultras, NICOLAS-GABRIEL GALOIS had been a popular mayor. The inhabitants went out to meet the hearse on its arrival from Paris, as far as the nearby village of Bagneux, where some of them took the coffin and carried it on their shoulders as far as their parish church. The new parish priest was there to take the funeral service before burial, but was met with insults and stones, one of which hit him on the forehead.

This scene took place in the presence of EVARISTE, who was grief

Fig. 8: Tomb of NICOLAS-GABRIEL GALOIS in the cemetery of Bourg-la-Reine

stricken and shocked by the recent hate campaign waged against his father, which had been the cause of his death. He immediately realized that the political clash that had taken place by the grave ready for his father's body had been the result of political fervour that would, from that moment on, disturb his spirit as well.

The entire population of Bourg-la-Reine contributed to the cost of a funerary monument, on which the following epitaph was engraved:

> Like an angel of peace
> sent to earth every
> day where he walked, he cast
> good things and takes with him to the grave
> the everlasting grief of the unhappy men
> who called him father.
>
> If all those who knew him grieve him
> how great must the grief of his loved ones be?
> Who will be able to console his children and his widow
> for the horrible blow with which
> your death has struck me?
> My love, my hope, my rest
> all is here.
> *May he rest in peace.*

There is still, to this day, a large plaque in memory of NICOLAS-GABRIEL GALOIS, Mayor for fifteen years, on the façade of Bourg-la-Reine Town Hall.

NATHALIE-THÉODORE, EVARISTE's sister, who had recently become Mme CHANTELOT did not need a guardian, but someone had to take responsibility for EVARISTE and ALFRED. Ten days after NICOLAS-GABRIEL's tragic death, on 12 July, the family gathered together and chose THÉODORE-MICHEL, the retired lieutenant-colonel as the boys' guardian. Nobody, though, would ever have been able to take his father's place in EVARISTE's heart.

Before that terrible July of 1829 was over, another event was due that would cast the still suffering GALOIS into a state of the blackest depression. The moment had come to take the entrance examination to the *Ecole Polytechnique*, for the second time. Unfortunately, as was pointed out twenty years later, "an examination candidate of superior intelligence is lost if he finds an examiner of inferior intelligence." The only claim to fame of DINET, one of the two examiners, is that he failed GALOIS at this examination. The second examiner, LEFÉBURE DE FOURCY, was the author of a num-

Fig. 9: Commemorative plaque of Mayor NICOLAS-GABRIEL GALOIS on the façade of Bourg-la-Reine Town Hall

ber of boring school textbooks, which are probably still gathering dust in some library.

This examination has become something of a legend in the history of mathematics. GALOIS was asked to describe the theory of logarithms. He did not keep to the traditional textbooks and was criticized by the examiners. A heated discussion began, during which GALOIS, who was sure he was right, threw the backboard duster at one of them. Needless to say, he was not admitted. Since the examination could only be taken twice, his greatest dream had now faded away for ever.

No longer being able to attend the *Ecole Polytechnique*, GALOIS now had to face the problem of what to do about his education. The situation was made worse by the poor financial resources of his family after his father's death. He needed financial assistance. One way of getting a grant was to attend the *Ecole Préparatoire*.

The course there lasted for two years, and trained students to become school teachers. Founded during the Revolution, on the basis of a project by LAKANAL, the *Ecole Normale*, its predecessor, had previously been responsible for teacher training and had numbered many excellent scholars among its teachers, such as LA-

GRANGE, MONGE, BERNADIN DE SAINT-PIERRE and BERTHOLLET the chemist. It had been closed down in 1822 and replaced, on the initiative of Mons. DENIS-LUC FRAYSSINOUS, by the *Ecole Préparatoire*, which was housed in the *Collège de Plessis*, next to the Hôtel de Langres. Its first director had been ARMAND MARRAST, who was soon followed by JOSEPH-DANIEL GUIGNIAULT, a Greek literature specialist, more acceptable to the legitimist party, which had left its imprint on the organization of the institute. As far as possible, the students were also chosen on the basis of their political opinions.

Though officially secular, the *Ecole Préparatoire* imposed a host of religious practices on its students. Their day began with prayers in assembly, and there were also prayers at mealtimes. It was compulsory to read something morally uplifting in the evening, and confession was required at least once per month. Failure to go to confession for two months running was enough to be expelled.

As can be imagined, GALOIS was not at all enthusiastic about going to the *Ecole Préparatoire*, but he had no real choice and, though the last date for presentation of the application to take the entrance examination had expired, it was with the help, once again, of the kind RICHARD that he managed to take it nonetheless.

On 12 August, a letter was passed on by ROUSSELLE, the secretary of the Paris Academy, to the secretary of the *Ecole*, in which EVARISTE GALOIS stated that encouragement from scientists accompanied his own wish to embark on a teaching career. Thus, despite this technical irregularity, he was allowed to take the entrance examination written papers on 20 and 25 August. The examining board was made up of the mathematician CHARLES-ANTOINE-FRANÇOIS LEROY, the physicist JEAN-CLAUDE-EUGÈNE PÉCLET and general inspector DE COURVILLE. GALOIS came second out of five candidates.

Nevertheless, on 31 August, GALOIS was still uncertain about his decision and in a very pessimistic frame of mind. In a letter to his uncle ANTOINE DEMANTE, his mother's brother, he wrote:

> Here I am again in doubt about my career. What saddens me is that this indecision is more likely to decrease than increase my options. Hesitating and worrying are not very pleasant situations to be in, but it is very easy not to find pleasant situations where I am.

Definite admission to the *Ecole*, however, required candidates to take a diploma (or *baccalaureate*) in arts and sciences and sit for an oral examination. GALOIS' total lack of interest in any sub-

Fig. 10: General view of the building containing the *Lycée Louis-le-Grand* and the *Ecole Normale*, seen from rue St-Jacques

ject that was not mathematics made it very difficult for him to take these examinations. He had a first try at the *baccalaureate* in arts on 9 December and failed. Only one week later, the examining board comprising LECLERC, BARBIÉ DU BOCAGE, VICTOR COUSIN and THILLAYE, decided to accept his literary work, which was actually pretty poor. On 29 December, FRANCŒUR, HACHETTE, and LEFÉBURE DE FOURCY awarded him his *baccalaureate* in science.

It was only thanks to his good results in the oral examination in mathematics that GALOIS was finally admitted to the *Ecole Préparatoire*. LEROY's opinion was, on the whole, positive, though he only gave him 8 out of 10:

> This candidate has some obscure areas in the expression of his ideas, but he is intelligent and shows great aptitude for research. He provided me with a fresh view of applied analysis.

His physics examination was so bad that PÉCLET wrote in his report:

> He is the only candidate who gave poor answers. He knows nothing. I have been told that he is good at mathematics. This greatly surprises me, since, to judge by the examination, he does not appear to be very intelligent, or

perhaps his intelligence is so well hidden that I was unable to uncover it. If he really is what he appears to be, I doubt very much that he will make a good teacher.

GALOIS was accepted, nonetheless, and, on 20 February 1830, signed a pledge to remain within the state education system for ten years.

The Beginning of a Difficult Year

It is very likely that GALOIS had been reluctant to sit for these examinations, anxious as he was to know the result of the Academy's scrutiny of the *mémoires* he had presented the previous spring. The results had still not arrived. CAUCHY, who had taken the manuscripts home, had made no further mention of them until 18 January 1830, when he wrote a letter to the President to justify his absence from that day's session. He declared that it was during that session that he had intended to present his report on the work of the young GALOIS, and, therefore, asked the President to put this report on the agenda for the following session. The following week, CAUCHY was present, but, forgetting what he had previously written, presented some of his own research.

From that moment on, CAUCHY never again mentioned GALOIS, who waited in vain for his judgment, and was never able to get his manuscripts back. Only six months previously, after LEGENDRE had made the announcement of ABEL's death before the Academy, on 22 June 1829, CAUCHY had decided to prepare a report on the *mémoire* presented by ABEL, which he had been keeping among his own papers for the previous three years. This move may have been caused by a feeling of remorse over his neglect, or, more likely, by a plea from the Norwegian Consul.

GALOIS' attitude did not change when he started his new school. He continued to study nothing but mathematics, along his own lines, and did not even disguise his scorn for his teachers and his total lack of interest in other subjects.

Disappointed by CAUCHY's silence, he decided to try and interest the Academy in his research once more. As of 2 June, the Academy had announced the award of a prize (the *Grand Prix de Mathématiques*) to a paper, either printed or in manuscript, that included the most notable application of mathematical theories to general physics or astronomy, or *an important analytical discovery*. The examining board, however, was only set up in January 1830, consisting of LACROIX, POISSON, LEGENDRE and POINSOT. The expiry date for presentation of papers was 1 March.

This was a wonderful opportunity which had to be taken up. GALOIS rewrote his paper, perhaps with an occasional modification in respect of the previous version, and was the last to present it to the Academy secretariat. Readers familiar with mathematics will recognize the names of his rivals: JACOBI, LIBRI, PONCELET, STURM, LAMÉ, LIOUVILLE, PLANA, DIRICHLET. This, however, had no effect on GALOIS, who was sure of his originality, and the importance of his research. Neither was he disturbed by the fact that the board, following the regulations, was allowed to consider work published between 1 January 1828 and 1 January 1830, even without the authors having officially applied, which could also include the by now famous name of ABEL.

Bad luck seems to have decided to continue tormenting GALOIS. FOURIER, without any real justification, took his manuscript home, and a few weeks later, on 16 May, died. GALOIS' paper was never found and its author, with hardly any difficulty, was excluded from the prize. Nobody took the trouble of informing him.

Nevertheless, at this stage it is worthwhile deflating the legend of an entirely misunderstood GALOIS, which some of his biographers (E. T. BELL, in particular) have created. Actually, in the spring of 1830, he had the satisfaction of seeing three short articles of his published in the prestigious *Bulletin de Férussac*, a review that only published work by established scientists. In the April number a short, but anything but elementary, article appeared on the algebraic solution of equations. There were seven articles in this issue, by, apart from GALOIS, such scholars as GERGONNE, CHASLES, JACOBI, POISSON, LIBRI and the great CAUCHY. Two further short articles, one on the solution of numerical equations, and the other on the theory of numbers were published in the June issue. Together with GALOIS' contributions, there were others by CAUCHY, CHASLES and POISSON.

GALOIS had been prevented from making friends with his school fellows at the *Louis-le-Grand* by his rebellious character and awareness of intellectual superiority. At the *Ecole Préparatoire*, he did, at last, find a friend in a student from the second year called AUGUSTE CHEVALIER. AUGUSTE and his brother MICHEL, who was a student at the *Ecole Polytechnique* at the time, and was to become a famous economist, had a great influence on GALOIS, whose interest in politics was aroused by long, heated discussions with his

Fig. 11: Montagne Ste Geneviève seen from the *Ecole Polytechnique*. In the background the church of St-Etienne-du-Mont and the Panthéon can be seen.

new friends. At the time, the CHEVALIER brothers were followers of Saint-Simonianism, the utopian theory deriving from the ideas of CLAUDE-HENRI DE ROUVROY, comte DE SAINT-SIMON.

SAINT-SIMON was born in 1760, and is one of the most interesting people who lived in the late eighteenth and early nineteenth centuries. He sailed to America, when only nineteen, fighting courageously in the army sent by France to support the American rebels. His reforming zeal seems to date from that period. During the French Revolution, he gave up his title, taking the name of Citizen BONHOMME. He believed that the necessary conditions for progress consisted in the elimination of poverty and ignorance. This could be achieved by turning Christianity into a secular religion, and the granting of spiritual power to scientists rather than priests. When he died in 1826, his most loyal followers set up an association for the further spread of his teachings. Between 1826 and 1829 they managed to find many adherents at the *Ecole Polytechnique*. At first, the meetings of the association were held in the Paris apartment of HIPPOLYTE and SADI CARNOT (the latter being a famous physicist), the sons of the renowned mathematician and politician LAZARE CARNOT. One of their first moves was the publication of the newspaper called *Le Producteur*, which only lasted one year. One of its most frequent contributors was the philosopher COMTE.

However, the best known Saint-Simonian newspaper was *Le Globe*, founded by PIERRE LEROUX in 1824, and among its early editorial staff were LEROUX himself and HIPPOLYTE CARNOT. It was in *Le Globe* that the word "socialism" appeared in print for the first time, in February 1832.

Saint-Simonianism soon began to take on the character of a religious sect under the guidance of PROSPER ENFANTIN, who was appropriately addressed as *père*. From 1830 onwards, the members of the sect provided the first statement of their beliefs in *Exposition de la doctrine de Saint-Simon*. These were based on two basic ideas: public ownership and the abolition of social inequality:

> Here is the new right that takes the place of that of conquest and birth: man will no longer exploit other men, but man in association with other men, will exploit the world subjected to his strength... We have reached the conclusion that the future to which [the human race] is moving is a condition in which all strength will be combined for peaceful ends.

SAINT-SIMON's intentions were surpassed by his followers, who also occupied themselves with problems such as women's emancipation, and improvement in the treatment of the mentally ill and criminals. The doctrine, however, changed considerably as it developed.

I have been unable to discover to what extent GALOIS accepted Saint-Simonianism. It is certain that the CHEVALIER brothers did influence him considerably in his interest in politics, which even caused him to neglect his mathematics. On 22 June 1830, GALOIS took the first year final examinations at the *Ecole Préparatoire*. The subject was differential and integral calculus. The examining board consisted of CAUCHY, HACHETTE, and LEFÉBURE DE FOURCY. GALOIS passed, but was only fourth out of eight candidates, being given 8 marks like his fellow candidate POLLET. Both had been beaten by PINARD and LASSASSAIGNE, who were given 10 and 9 marks respectively.

3 The "Three Glorious Days"

The Ordinances of CHARLES X

MARTIGNAC's fall from power was brought about by his attempt to make France even a slightly more democratic country, by abolishing government nomination of the members of certain local councils, and making them elective, even by a limited electorate. His project was attacked by both right and left, and this general discontent favoured the King's intrigues with the various legitimist factions aiming at the creation of a council of Ministers which was wholly acceptable to him.

The completion of a new government, made up of Ultras, was announced on 8 August 1829. Prince JULES DE POLIGNAC, an ex *émigré* and active member of the *Congrégation* was chosen as President of the Council of Ministers. Comte DE LA BOURDONNAIE, one of the leading adherents of the White Terror, became Minister of the Interior. General BOURMONT, who had betrayed NAPOLEON on the eve of Waterloo, was chosen as Minister of War.

The opponents of the Bourbons quickly formed two new parties; one of them republican and the other Orleanist. The first group, which was mostly made up of young workers and students, acknowledged GODEFROY CAVAIGNAC, GARNIER-PAGÈS and HIPPOLYTE CARNOT as their leaders, and had *La Tribune* newspaper as their means of political propaganda. The second group, led by the banker LAFITTE, included those who wanted the young duc D'ORLÉANS, LOUIS-PHILIPPE, on the throne, in place of CHARLES X. The Orleanist pretender was the son of PHILIPPE EGALITÉ, who had fought in the battles of Valmy and Jemappes during the Revolution, which he supported. The journalist THIERS, author of a very popular *Histoire de la Révolution française*, was one of the main supporters of the Orleanists. In January 1830, he had started up *Le National* newspaper, as an organ of party propaganda.

The new government waited until March 1830 to call Parlia-

Fig. 12: The duc D'ORLÉANS as a young man

ment, after seven idle months. The speech from the throne, at the Opening of Parliament, contained a veiled threat of a *coup d'état*. The contrary vote of the deputies, in support of the Constitution, was 221 against 182. CHARLES X reacted by dissolving Parliament and new elections were held at the beginning of July. The results were: 274 seats for the opposition and 143 for the supporters of the government. Nonetheless, the King refused to accept the decision of the electorate, and, by means of an extremely personal interpretation of article 14 of the Constitution, which gave him the power to issue "the necessary ordinances for the enforcement of law and the security of the State," issued four such ordinances, which were actually nothing more than a *coup d'état*. The first ordinance concerned the freedom of the press and read as follows:

> Freedom of the press is suspended ... consequently no newspaper or pamphlet, of a periodical or semi-periodical nature, already in existence or planned, without reference to the subjects treated, may appear in Paris, or in the departments, without authorization separately obtained from us by authors and printers.
> This authorization must be renewed at intervals of three months.
> It may be revoked.

The first article of the second ordinance provided for the dissolution of the Chamber of Deputies, thus annulling the elections.

The third ordinance contained the following:

> The property qualification for voters and candidates shall consist only of the sums for which the electors and candidates are personally enrolled, either as owners or life tenants, in the property tax or personal tax registers.

A new electoral law was thus enacted, consisting of 30 detailed articles.

The last ordinance set the dates for the new elections: 6 September for the convocation of local electoral colleges, and 18 September for those of the departments.

On the morning of 26 July 1830, the newspaper *Le Moniteur*, which was publishing the King's ordinances, only appeared between 9.00 and 10.00 a.m., later than usual, owing to the complicated nature and importance of the texts to be prepared for printing. The paper also contained a report by CHANTELAUZE, Keeper of the Seals, and a decree from MANGIN, the prefect of police, which read as follows:

> All owners of reading rooms, cafés etc., who permit the public to read newspapers and other printed matter violating the ordinance of the 26 day

of this month, regarding the press, will be prosecuted as accomplices of the crimes of which the said printed matter may be accused and their premises will be subject to temporary closure.

Riots in Paris

In defiance of MANGIN's decree, four newspapers, *Le National, Le Globe, Le Temps* and *Le Journal de Commerce* appeared on the morning of 27 July, without the necessary royal authorization. They all had an article of protest written by THIERS the previous evening, and signed by 44 journalists representing the entire French press, on their front pages. The article incited the people to rebel:

> ... the government has violated legality. We are not obliged to obey [...] The government has today lost that character of legality which requires obedience. We shall resist for all that concerns us. It is France that must judge to what lengths resistance is to go.

Le National and *Le Globe* listed the names of the signatories. The other two newspapers were more cautious, only mentioning their number. Besides, CHARLES-FRANÇOIS RÉMUSAT appealed to the citizens from the columns of *Le Globe*:

> A crime has been committed. Ministers have obtained tyrannical ordinances from the King. We will only give in to violence. With all our strength, we invoke hatred against Polignac [...] The ordinances are null and void. We fearlessly entrust the most courageous nation in the world with the defence of freedom within the law.

La Gazette de France, La Quotidienne, and *L'Universal* were published after obtaining royal authorization. The other Parisian newspapers, cautiously, preferred not to appear at all. In the early morning, the rebel newspapers, in particular *Le National* and *Le Temps*, were distributed free.

An ever growing crowd began to fill the streets of Paris. Discussion was rife and cries of "Long Live *La Charte*" could be heard. At 11 a.m. two high ranking police officers, leading a troop of soldiers, arrived at the offices of *Le National*, in rue Neuve-Saint-Marc, with the order to confiscate all the copies of the newspaper. Unfortunately they found nothing to confiscate, because all the 7,000 copies had already been given away. The police, in anger and frustration, then set about smashing up the printing presses. There were no difficulties at *Le Journal de Commerce*, however, while nobody appeared at the offices of *Le Globe*, where RÉMUSAT, the editor, was impatiently awaiting the opportunity to stage a gesture of defiance.

The first serious rioting began in front of the Palais-Royal. The Prefect had ordered the local police chief to move people from the gardens and close the gates. The crowd then flowed into the square and neighbouring streets, but would not disperse. The shops in the area were quickly shut. The first warning rifle shots were soon heard. Nothing special happened, however, until the early hours of the afternoon, when a group of working class Parisians entered an area in the process of demolition, near the Galerie de Nemours, opposite the Café de la Regense, climbed up onto a pile of stones and began throwing them at the gendarmes, who still managed to empty the Palais-Royal area. Stones flew more thickly than before.

At 3.00 p.m., there were the first deaths. An Englishman, by the name of Fox, began firing from the balcony of the Hôtel Royal, at the corner of rue Saint-Honoré and rue des Pyramides. The nervous troops below returned his fire, without having received orders to shoot. The Englishman and his two servants were killed. From that moment, serious shooting began, causing more dead and wounded. The unruly crowd began to panic. Although it had initially been confined to the narrow streets around the Palais-Royal, rue de Richelieu, rue de Valois, rue de Fromenteau, rue de Chartres, the riot now began to spread to other parts of Paris. When a man placed the dead body of a girl, who had been hit by a bullet in the forehead, at the foot of the statue of LOUIS XIV, his cry of "Revenge, Revenge!", passing from mouth to mouth, began to ring round the whole city.

The royal guards were soon posted to protect the centres of power. The 7th Regiment, with six cannon, was sent to place LOUIS XV, and the 1st Regiment, with two cannon and fifty lancers, was given the job of defending the Ministry of Foreign Affairs in boulevard des Capucines. The 3rd Regiment, with four cannon and crack gendarmes, was sent to the Carousel. The army, under the command of Marshal MARMONT now faced the people: the 15th Light Artillery, at the Pont Neuf, the 5th Regiment in place Vendôme, the 53rd on the boulevards Poissonière and Saint-Denis, the 50th and 1st Cuirassiers guarding place de la Bastille.

The crowds now needed arms, and began to pillage gunsmiths' shops. The first barricade, made up of three overturned carriages, was set up at the crossing of rue Saint-Honoré and rues de Richelieu and de Rohan. News of the revolt soon reached the Left Bank, and two companies of the 15th Regiment were sent to keep the students of the Latin Quarter under control. The *Ecole Polytechnique* was in ferment. CHARRAS, a former student, who had been expelled five months previously for singing the *Marseillaise* during a banquet, had made sure his colleagues were given copies of the newspapers with the journalists' statement of protest. BIRET, director of studies, forbad the students to assemble, but they did so, nevertheless, in the billiards room. A decision was quickly taken to go to the aid of the rebels, and a group made up of LOTHON, BERTHELIN, PINSONNIER and TOURNEAUX was sent ahead to explore. After dark the four students climbed over the outer wall, and ran to CHARRAS' home in rue des Fossés-du-Temple. Meanwhile, inside the *Ecole*, the students broke open the door of the fencing room, took all the rapiers available, and began to sharpen them on the tiles in the corridors.

When the printing presses had been put back into working order at *Le National*, the journalists came to an agreement with their colleagues at *Le Globe* to continue publishing a newspaper of some kind. In the evening LOUIS-AUGUSTE BLANQUI wrote an appeal to the people:

> Citizens of Paris!
>
> Charles X has torn up *La Charte*, overturned the laws, destroyed all freedom. We have no press! We have no newspapers! No books! The Chamber of Deputies has gone! The *ancien régime* has been restored, and France has been delivered, with her hands and feet bound, to the nobles and the priests.
>
> To arms, Citizens! To arms, to defend our country, our honour, our existence! Shall we let ourselves become a bunch of slaves under the whip of the Jesuits? No, no! We'd rather die!
>
> But the people will not die. Those who are trying to enslave them will die! Stand up! Stand up! Let us annihilate these evils! May retribution fall with the speed of lightening on their *coup d'état*.

There followed seven articles regulating the call to arms, of all citizens between the ages of sixteen and fifty, in defence of the Nation and freedom.

At dawn the following morning, the four students from the *Polytechnique*, after spending the night in CHARRAS' home, decided to call on MARTELET, their mathematics teacher. Knowing his political views, they were certain of getting a good reception. They hoped that they would be able to change into civilian clothes, which would be more comfortable than their uniforms, at his home, and then return to give their companions the signal to follow them. Other groups were also meeting to decide on what strategy to follow. The republicans met at *La Petite Jacobinière* bookshop, to get their orders from CAVAIGNAC. The Bonapartists, who had already met in General GOURGAUD's house the day before, gathered in place des Petits-Pères.

Arms were still needed. Two thousand swords and knives were found, and immediately seized from a knife maker's shop in rue Saint-Martin. Though harder to handle than guns, they were still useful to the cause of freedom!

Work continued at the offices of *Le National* and *Le Globe*, during the night, and both papers came out with communiqués, which were on display in the streets and on the tree trunks of the

Fig. 13: *Ecole Polytechnique* students' uniform in 1830

boulevards. When the signal was given by the four students, all their companions, armed with rapiers, left the *Ecole Polytechnique*, most of them proudly wearing their full dress uniforms. The words of their song could soon be heard echoing around the street up the Montagne-Sainte-Geneviève:

> Long live the youth of the college
> They never had more than one party:
> when Paris was under siege
> they were there, like today.
>
> Fellow Frenchmen, let us sing of the heroic courage
> of the youth of the *Ecole Polytechnique*
> They never knew danger
> to have revenge
> Frenchmen, to have revenge.

The people soon joined in with shouts of "Long Live the *Ecole Polytechnique!*", "Long Live *La Charte!*", "Long Live Freedom!".

A group of working men coming down the faubourg Saint-Marceau began to follow the students. One of the students tore off the white rosette from his two-cornered hat, threw it to the ground, stamped on it, and shouted "Down with the Bourbons!". All his companions followed his example, shouting: "Down with the Bourbons!".

The songs and shouts could also be heard in the rue Saint-Jacques, where GALOIS and his fellow students, in desperation, were forced to follow events from behind the barred windows. At 5.30 a.m., GUIGNIAULT, the director, had called his students together and ordered them not to leave the building, reminding them of their pledge of ten years service to the State, which already made them civil servants. They should be cautious, he told them, since they were privileged, unlike soldiers who were forced to choose between freedom and their binding oath of loyalty to the King.

GALOIS disliked these words intensely, and, together with his fellow student, BÉNARD, decided to disobey. His duty was to rush to the barricades. GUIGNIAULT's response was firm: he would call in the troops, if necessary, to stop his students mixing with the rebellious crowds in the streets. For the moment he would be content with barring all the doors giving onto the street. GALOIS now had another reason, and it was an important one, to envy the students of the *Ecole Polytechnique*.

In the meantime, the crowds were moving towards the Hôtel de Ville and, once they had arrived, started pushing on the doors, which soon fell in. The people of Paris had taken over one of the centres of power, which again became a symbol of freedom and popular sovereignty.

The white flag of the Bourbons had soon been taken down and thrown into the square below, where it was trodden on and torn to pieces. A collection was organized to buy two tricolours in rue de la Juiverie, one for the Hôtel de Ville, and another for Notre Dame. It was about 11.00 a.m., when the two flags could be seen flying. The crowd cheered, shouted, and wept with joy. All of a sudden, HIPPOLYTE DANIEL, a medical student from the Hôtel Dieu Hospital, left his duties, crossed the square and ran to begin ringing the great bell of Notre Dame. The bells of Saint-Séverin rang out in reply, almost immediately, and, soon after, all the church bells in Paris followed suit. It sounded like a signal. Almost as if by miracle, the streets of Paris were filled with tricolour flags, and echoed with the singing of the *Marseillaise*. At this point the people of Paris were thought to have sufficiently shown their strength, and it was agreed to send a delegate, in the person of the astronomer FRANÇOIS ARAGO, to the Tuileries, with the purpose of persuading MARMONT to abandon the city and go to the King at Saint-Cloud. He might have withdrawn the ordinances. Since he was afraid of being seen as a traitor to the people, by those who saw him enter the royal Palace, ARAGO took his little son with him. "A father," he said, "does not give his son lessons in treason." At the Tuileries he was immediately surrounded, threateningly, by the aides-de-camp. What was this scientist doing there on a day like this? He had difficulty in persuading them to take him to their commanding officer. His efforts were useless. MARMONT would not listen. How could a soldier flee in the face of civilians armed with stones and clubs? The battle went on throughout the day.

In the evening, GALOIS decided to act. Being forced to stay idle was a painful, humiliating experience for him. Trying not to be noticed by his school fellows, he went out into the courtyard and tried several times to climb the outer wall, and jump down into rue du Cimetière Saint-Benoit. He grazed his hands and knees, but each attempt failed. He was a useless prisoner, while the people of Paris were making history. This attempt, however, was his first political act.

At dawn on 29 July, six thousand barricades blocked the streets of Paris. Everything available was dragged into the streets to build them: tables, chests-of-drawers, beds, tree trunks, ladders, and all kinds of household goods.

Fig. 14: The astronomer FRANÇOIS ARAGO (1786–1853). He wrote a number of scientific treatises and an introduction to astronomy for the general reader. He was also active in politics. He made an important contribution to the abolition of slavery in the French colonies.

The weather was, like the previous days, going to be very hot, and this meant that the corpses had to be carted away. The idea was to carry them to the banks of the Seine, and, from there take them by boat to the Champs de Mars and pont de Grenelle. Since road transport was prevented from moving, provisions were delivered on foot.

A large number of soldiers had taken off their uniforms and joined the people in revolt. In the Latin Quarter it was the students of the *Polytechnique* who were in charge, among them CHARRAS, who had put his uniform back on. Continual desertions forced General MARTINOT and his troops to leave Paris. By midday, the rebels were masters of the capital. The price of the revolution had been 2,000 dead in the King's army, and 1,800 among the citizens of Paris, who had fallen in the cause of freedom.

Another King!

The winners of the "Three glorious Days," as the three days during which the citizens of Paris fought the King's army from the barricades are usually called, were mostly Bonapartists and republicans. However, a republic in July 1830 was still unthinkable, since the republicans, who were not yet well organized, did not have enough political cohesion to impose their viewpoint.

After MARMONT and his troops had left the city, the situation was taken over by the banker LAFITTE and CASIMIR PERIER. They persuaded LAFAYETTE to accept command of the National Guard.

CHARLES X, now isolated at Saint-Cloud, was unable to regain power and make the necessary concessions to the people.

The morning of 30 July was a decisive day in the history of France, and in GALOIS' life. The walls of Paris were covered with copies of a manifesto published by *Le National* written by THIERS, which put forward the name of the duc D'ORLÉANS as the new King:

> Charles X cannot return to Paris. He has shed the people's blood. A republic would bring with it a dangerous split. It would put us in a difficult position with the rest of Europe.
>
> The duc d'Orléans is a prince who is dedicated to the cause of the Revolution. The duc d'Orléans has never fought against us. The duc d'Orléans was at Jemmapes. The duc d'Orléans is a citizen King [...] The duc d'Orléans is making no statements. He is waiting for us to express our wishes. Let us express these wishes then, and he will accept *La Charte*, as we have always expected. He will wear the crown of the French people.

This is what could be read in all the streets of Paris. In all the

Fig. 15: Portrait of General LAFAYETTE as a young man

newspapers, on the other hand, readers found out that GUIGNIAULT, the director of the *Ecole Préparatoire*, offered the services of his students to the provisional government. GALOIS was disgusted by this hypocrisy. After refusing to support the rebels in the streets,

Fig. 16: The entry of the duc D'ORLÉANS into Paris on 30 July 1830

in the hour of danger, and preventing his students from doing the same, he was now making capital out of the courage of those who had died on the barricades. GALOIS swore to himself that he would unmask the man who had now become the symbol of cowardice and opportunism.

On the night of 30 July, the duc D'ORLÉANS, who had sought refuge at Raincy, entered Paris and accepted the temporary regency offered to him by a delegation from the Chamber of Deputies, on the morning of 31. In the afternoon, he went to the Hôtel de Ville to receive the people's investiture. The people of Paris were witnesses to the affectionate meeting between the Duke and LAFAYETTE, which they greeted with cheers.

On 9 August, the Coronation, following a new, simpler rite, took place. The new King took the title of LOUIS-PHILIPPE I "King of the French." On 28 August, the King reviewed the National Guard, presenting LAFAYETTE with the new tricolour banners. They were decorated with silver-coloured fringes, and bore the words: *Liberté, égalité, ordre public, 27, 28, 29 juillet 1830*. Thus the republicans saw their victory transformed into a bitter defeat, but, as CAVAIGNAC said: "We only made concessions, because we were too weak."

After abdicating in favour of his grandson, the duc DE BORDEAUX, the son born to the duc DE BERRY's widow, CHARLES X left for exile, first in Lullworth in England, and then at Holyrood House in Edinburgh. As a demonstration of loyalty to the Bourbons, the mathematician CAUCHY also left France. He might have altered EVARISTE GALOIS' destiny. Science will never be able to forgive him for his negligence.

GALOIS returned to Bourg-la-Reine for the summer holidays of 1830. This was to be his last long stay with his family. As of the autumn of that year, he was to see them less and less. His mother, and his brother and sister noticed how much he had changed. Once shy and retiring, he was now bold and resolute. His family were astonished by his violent speeches on the rights of the masses, on how the July revolutionaries had been betrayed and why a second violent uprising was now necessary in Paris, as soon as possible. He declared that he was ready to fight, and even sacrifice himself for the republican cause, and would proudly repeat: "If I were only sure that a body would be enough to incite the people to revolt, I would offer mine."

At the beginning of October, he returned to Paris for his second year at the institution that, on 6 August, had revived its previous name of *Ecole Normale*. His friendship with the CHEVALIER brothers, especially AUGUSTE, was consolidated, but he also sought out young republicans, with whom he felt a certain affinity of ideals. They included: AUGUSTE BLANQUI, FRANÇOIS-VINCENT RASPAIL and NAPOLÉON-AIMÉ LEBON. GALOIS' new friends were beginning those activities that were to transform them into some of the most illustrious Frenchmen of the 19th century.

Blanquism, a socialist theory aiming at the establishment of a dictatorship of the proletariat, which would confiscate the goods of the rich and give rise to total equality, was still to be elaborated. BLANQUI was only twenty seven. He was a journalist with *Le Globe*, and a left wing Jacobin, who adhered to BABEUF's egalitarian tradition.

RASPAIL was ten years older, and split his time between political activity and natural sciences. German scholars had already acknowledged the importance of his work on the classification of grasses and, in that momentous year of 1830, he had published an interesting treatise entitled: "Essai de Chimie microscopique appliqué à la physiologie."

Wishing to contribute more than mere ideological support to the republican cause, GALOIS, BLANQUI, RASPAIL and LEBON joined the

Société des Amis du Peuple, which had been founded that very year and consisted of the most active and aggressive members of the republican party. The Society had not started life as a secret organization, and its members opposed the application of article 291 of the penal code, which allowed the authorities to forbid public meetings of more than twenty people. There was no mystery involved with becoming a member. Members were accepted because they were already known or on the basis of a declaration of patriotism. Being a patriot did not necessarily mean being a republican, though the majority of members were engaged in planning a republican revolution. The leaders of the Society were particularly anxious to attract influential or talented men, so as to be able to use their names or writings for propaganda purposes. Patriots were zealously prepared for the time when the right moment would come for an uprising. The first headquarters of the Society, which was presided over by JEAN-LOUIS HUBERT, was the PELTIER riding school in rue Montmartre. During meetings, that were open to the public and announced in advertisements in newspapers and by street posters, so as to get round article 291, twenty members sat in the central ring, while the others, in the guise of spectators, sat around the ring.

The Friends of the People also had their own newspaper, which was distributed in limited numbers, but which assisted the spread of news, thanks to an unusual device: the headlines were deliberately very long, so that the newspaper sellers ended up by shouting out most of the contents of the various articles.

The government press described the Friends of the People as particularly dangerous individuals, as far as the safety of individual citizens was concerned, since, it was claimed that they were ready to use any violent means to obtain what they wanted. Law abiding citizens tended to be rather frightened by all this, and it was not unusual to see shopkeepers quickly pulling the shutters of their shops down, when a well-known member of the Society walked past. When GALOIS became a member, probably in mid-November, the Society had entered a new phase, following the first political trial of the reign of LOUIS-PHILIPPE.

At the end of September, the members of the society had decided to provoke public debate on the legality of the Chamber elected before the July Revolution, and which the new King had confirmed. The discussion kept the citizens of Paris occupied, as they recalled the events that had seen them on the barricades for three whole days. In the end, it was decided that the deputies' mandate should be considered to have run out and that the people had the right to expect new elections.

A poster to be displayed throughout Paris was approved by those present in the PELTIER riding school and delivered to the printer DAVID. The handwritten version was, however, confiscated by the police and HUBERT and DAVID arrested.

The trial of the two men ended in HUBERT being given a prison sentence of three months and the printer being found not guilty. The court decreed the suppression of the *Société des Amis du Peuple*, which from that moment was forced to become a secret association. RASPAIL became the new president. The headquarters was transferred to rue Grenelle Saint-Honoré, and the meetings, which were no longer called through newspaper advertisements or posters were no longer public. Conditions for membership also became stricter.

The Society was far from being defunct. Its propaganda became more intense, and it started up an armed organization under the cover of the artillery of the National Guard. Of the four batteries of which it was composed, the second was commanded by GUINARD and CAVAIGNAC, and the third by BASTIDE and THOMAS, all influential members of the Society, and both were made up of members of the republican party.

The National Guard had come together spontaneously after 14 July 1789, and had been made official by the Constituent Assembly. It was a military organization of a very special kind, quite different from the rest of the French army. It had its own uniform, its banners, its own music, and fanfares but, above all, the usual military discipline was missing.

The End of a School Career

The first months of LOUIS-PHILIPPE's reign had witnessed the rise to influential positions of ambiguous individuals, who, with great ability, had managed to take advantage of the political changes. Among them was the philosopher VICTOR COUSIN, who, on 28 July, had said to BLANQUI "Sir, your colours may be the tricolour, but they will never be mine. The flag of France is a white flag!". During the previous reign, COUSIN had frequently asked for the Royal Public Education Board to be abolished, and now he was a member of it. He was also in charge of philosophy teaching in the *lycées*, and supervised the maintenance of standards at the *Ecole Normale*.

The "Three Glorious Days"

Fig. 17: GODEFROY CAVAIGNAC in the uniform of the National Guard

It was not only emotionally engaged youngsters such as GALOIS who were scandalized by such behaviour; the same was true of more moderate opinion. In a letter dated 16 October EUGÈNE BURNOUF, at the time a teacher at the *Ecole Normale*, wrote the following to the orientalist JULIUS MOHL:

> We are governed by [those who], at great length, discuss law and order, dangerous utopias, agitators etc. Then, cleverly taking over positions left vacant by the Ultras, they, surreptitiously, piece together conditions under Charles X ...
>
> But in the midst of all this, do not weep over the fate of our friend Cousin: he is well established on the Royal Board ...

Under the protection of his friend COUSIN, GUIGNIAULT reorganized the *Ecole Normale* in accordance with the basic principle that "A good student is not interested in politics." GALOIS, who did not hide his republican convictions, but had emphatically explained the party programme to his fellow students, soon found himself in conflict with the director, who would have given anything not to have such a subversive young man among his students.

GALOIS continually did things that irritated GUIGNIAULT. He asked if the students could wear a uniform like those at the *Ecole Polytechnique*, and the latter refused. A few days later, he asked if they could be armed, so that they could do military training. GUIGNIAULT thought the request ridiculous.

He was also always criticizing the organization of courses, which had been modified by the director, making the whole course last three instead of the previous two years. He was isolated by the other students, who were afraid of attracting the suspicion of the school authorities, if they made friends with him. He was even more isolated, when the director punished him by not allowing him out of the building for an indefinite period.

At the time, two newspapers, *Le Lycée* and *La Gazette des Ecoles* addressed themselves to a student readership. The two papers had different opinions and often attacked each other explicitly. The 2 December issue of *Le Lycée* published a letter from GUIGNIAULT attacking GUILLARD, a teacher at the *Louis-le-Grand*, who also worked for *La Gazette des Ecoles*. The answer came three days later, when *La Gazette* published an editorial on how GUIGNIAULT had progressed in his career:

> Instead, like cowards, of telling M. Guigniault that he cleverly took advantage of the illness of M. Gibbon, director of the *Ecole Préparatoire*, so as to take his place, we shall only be vague about the ambitious, and

The "Three Glorious Days"

intriguers. Instead of saying that M. Guigniault ... schemed to obtain the directorship of the school, and then the position of general inspector; instead of saying, by "conjecture" that he would be happy to be given the title of chief adviser to the *Ecole Normale*, if he were lucky enough to have things return to the old ways, we would prefer to criticize generally over-quick promotions ...; instead of saying that, not content with this, he goes out of his way to obtain all the little comforts in life, at the expense of others ... we will keep silent, because silence is more courteous.

The article ended with the announcement of a further contribution to the controversy:

> We could have no better conclusion to our reply than following it with a letter we have just received.

The letter, dated 3 December, came from an anonymous "student at the *Ecole Normale*":

> The letter from M. Guigniault in yesterday's *Lycée*, on the occasion of an article in your newspaper, seemed to be quite out of place. I thought you would be interested in any attempt to unmask this man.
>
> Here are the facts to which 46 students can testify. On the morning of 28 July, since many of the students wished to join in the uprising, M. Guigniault told them, twice, that he could could call the police to restore order. The police on 28 July!
>
> On the same day, he said to us, equally pedantically, "Many brave people have been killed on both sides. If I were a soldier, I would not know what decision to take. What should I sacrifice, freedom or legitimacy?"
>
> This is the man, who stuck a tricolour rosette on his hat the day after. These are our convinced liberals!
>
> I should also like to inform you, sir, that the students of the *Ecole Normale*, inspired by noble patriotic spirits, very recently presented themselves to M. Guigniault, to inform him of their intention to address a petition to the Ministry of Education, asking for arms, and wishing to take part in military training, so as to be able to defend their territory if required.
>
> Here is M. Guigniault's answer. It is liberal just like his answer of 28 July:
>
> "The request addressed to me would make us look ridiculous. It is an imitation of what has been done in higher level institutions: it came from below. I should like to point out that, when the same request reached the Minister from these institutions of higher education, only two members of the Royal Board voted in favour, and they were not among the liberals. The Minister accepted, because he feared the students' turbulence, which will be ruinous for the University and *Ecole Polytechnique*." I believe that, from one point of view, M. Guigniault is right to defend himself in this way, against being blamed for his prejudice against the new *Ecole Normale*. He only loves the old *Ecole Normale*, which had everything.
>
> We recently asked for a uniform, which was denied us; they did not

wear them at the old school. In the old school, the course lasted three years. Although, when the new school was set up, the third year was acknowledged to be pointless, M. Guigniault brought it back.

Soon, following the rules of the old *Ecole Normale*, we will only be allowed out once a month, and will have to return by 5.00 p.m. It is wonderful to belong to the educational system that produced men like Cousin and Guigniault!

Everything he does show his narrow outlook and ingrained conservatism.

Sir, I hope that these details will interest you, and that you will put them to the use you think fit, to the benefit of your excellent newspaper.

The newspaper added the following note:

We have removed the signature from this letter, although we were not requested to do so. We should also like to point out that, immediately after the Three Glorious Days, M. Guigniault sent a communiqué to all the newspapers that the director of the *Ecole Normal* offered the services of his students to the provisional government.

Was GALOIS the anonymous "student at the *Ecole Normale*"? When questioned by both the director and his fellow students, he did not confirm that he was the author, but neither did he deny it, showing total indifference to the consequences of his attitude. GUIGNIAULT and VICTOR COUSIN, in whose opinion the republicans were a disgrace to mankind, though having no proof of his guilt, were only too happy to take advantage of the occasion to rid themselves of an undesirable student, who could easily spread discontent among his fellows.

When, on 9 December, GUIGNIAULT had GALOIS' mother come for him, she was no longer living in Bourg-la-Reine, having moved to Paris, where, owing to difficult financial circumstances, she had been forced to accept a post as a lady's companion.

On the evening of the same day, the director communicated GALOIS' expulsion to the Minister, maintaining that the young man had fully confessed his guilt:

I deeply regret to inform you of a decision I have been forced to make, on my own responsibility, and of which I apply for immediate official approval. I have expelled Evariste Galois from the *Ecole Normale* [...]

This student has been shown, on the basis of a full confession of his impudence, to have been the author of a piece of writing that has provoked the indignation of the entire *Ecole*. The piece in question consists of a letter published in the *Gazette des Ecoles* [...] This letter appeared to me ... to have damaged the very reputation of the *Ecole Normale* to such an extent, that I could do nothing but follow the matter up [...]

The "Three Glorious Days" 73

> I have had reason to complain about Galois from the very moment he arrived ... nevertheless, in my concern for his undoubted talent for mathematics ... I tolerated his unconventional behaviour, his laziness and his very difficult character, in the hope that, though unable to alter his morality, I could enable him to complete his second year ... without causing grief to his mother, who, I knew, counted on her son's future for her own well being.

GALOIS' fellow students, in their worry about not losing the director's benevolence, which was so important for their future careers, also expressed their views, by writing a letter to *La Gazette*. Actually there were two letters: one from the arts students, and another from those studying science. The former adopted an opportunist stance as regards their director:

> ... we hasten to show our appreciation to M. Guigniault for the honourable and resolute manner in which he has defended our interests during the entire period of his directorship, and at the most critical moments in the life of our institution. We declare that it is thanks to him that we enjoyed the freedom of thought, which everywhere else was being suffocated, and that, during the last days of July, his attitude towards us was coherent with his previous behaviour.

The students who signed this statement were: HAMEL, GUÉRARD, DUPREY, NENS-LAFAIST, ROUX, MONIN, HUGUENIN, BARY, DABAS, CAPELLE, COLLET, VENDEYES and DESMAROUX.

GALOIS had asked his fellow science students to show him their solidarity somehow. They refused, but they did so without taking GUIGNIAULT's side:

> The students ... not being witnesses to the facts, reject the appeal from the author of the letter published in the 5 December issue of *La Gazette des Ecoles* for the testimony he requires.

The students in question were: POLLET, LASSASSAIGNE, BISSEY, PINAUD, LAURENT, CHOFFER and GÉRARD.

La Gazette published the following statement, in the same number as that containing the two letters from students, as a contribution to the ongoing controversy:

> We have just heard that the director of the *Ecole Normale* ... summoned all his students to an assembly, at which, he asked each one individually the following question: "Are you the author of the letter to *La Gazette des Ecoles*?". The first four answered in the negative, while the fifth answered: "Sir, I do not think I can answer this question, because it would help in betraying one of my fellow students." M. Guigniault was extremely irritated by this proud, noble reply.

GALOIS' expulsion soon became general knowledge, and in the 12 December issue of *Le Constitutionel*, the editorial staff published an appeal to the Minister on his behalf:

> We believe that we should bring to the attention of the Minister of Education a misuse of power, of which one of the best students of the *Ecole Normal* has been a victim ... We hope that M. Mérilhon, who has already shown his good judgment in the face of controversy, will ask for an enquiry into the whole question.

The appeal went unheeded. Actually, GALOIS must have felt relieved that he would no longer be required to follow pointless lessons in a hostile atmosphere, where he was considered to be "an extremely negative figure, with a perverse, sly personality," as one of his fellow students wrote of him. He now had all the time he needed to devote himself to politics and go on with his mathematical research, in his own way.

His first move, in this new status, was to enlist in the artillery of the National Guard. He could at last proudly wear a real uniform and be armed! He was a soldier, who, if the need arose, would be able to fight for the freedom and glory of France, no longer an immature student, under the orders of a hypocritical director. As an artilleryman, he soon had the opportunity to see active service. On 15 December, the trial of CHARLES X's ministers began before the Chamber of Peers, which, for the occasion, had been turned into a high court. The people of Paris were agitated, and there were some who demanded capital punishment for the accused.

On 21 December, the Chamber of Peers issued a sentence of life imprisonment. On that day, GALOIS was stationed with his company in the courtyard of the Louvre. A revolt was feared, and, for reasons of security, the prisoners were being kept in the fortress at Vincennes.

Meanwhile, the arguments over GALOIS' expulsion from the *Ecole Normale* gave no sign of coming to an end. GUIGNIAULT stubbornly continued to sling mud at his former student. He wrote to the Minister again, on 14 December, including a letter:

> ... sent to the *Gazette* ... by one of our best students ... a bright young man of sound character, who, on occasion, deserved to be the recipient of my most secret thoughts, during the difficult times we experienced together.

The letter had been written by BACH, who had energetically defended the director's behaviour during the events of the previous July. The Ordinances, BACH maintained, had deeply disturbed

GUIGNIAULT, who had warned his students that he would have to sacrifice himself in the liberal cause, since: "... he could not have foreseen that three days later the people would punish the traitor and win its freedom."

It was the director's sense of responsibility towards parents that had persuaded him not to allow the students to join the rebels on the barricades. Far from threatening to call the police, he had only asked those who insisted on going out to wait until the following day, and not leave the building without advance warning. It was also false, BACH added, that GUIGNIAULT was supposed to have said:

> that, if he had been a soldier, he would not have known which way to turn. He simply acknowledged, like everybody else, that the soldiers were in an uncomfortable position, since they were in the dilemma of either sacrificing freedom or disobeying the oath that kept them in the army ...

BACH added that he had never heard the director use the word "legitimacy," in all the time he had been at school and concluded:

> There are as many lies as sentences. It is also false that, immediately after the victory, M. Guigniault hurriedly put an enormous tricolour rosette on his hat. Five students had been chosen to accompany the body of young Farcy, who had been killed during the Three Days, and I was among them. We went to the director, who was to join us in the cortège. We were all wearing the national colours. He was the only one without a rosette: "Gentlemen," he said, "you have done what I was going to do. It has been in our hearts for so long. Now we can also wear it here." You all know that he was speaking the truth.

Still dissatisfied, four days later, GUIGNIAULT wrote to *Le Constitutionel* again, highlighting the fact that the students of the *Ecole Normale* had all come to his defence.

It was GALOIS, however, who had the last word. On 30 December he got *La Gazette* to publish the following appeal to his ex school fellows:

> It is up to neither you nor me to pass final judgment on M. Guigniault's rights. But you must not allow one thing: that he attribute the whole responsibility for my expulsion to you, and that, after the brotherly affection you showed me when I left, he dares to claim that it was you who wanted me to be expelled [...] Dear friends, do something more. I am not asking anything for myself, but speak out for your own honour and according to your conscience.

4 To Louis-Philippe!

A New Plan

On 31 December 1830, GALOIS, together with all the other republicans, was to encounter another serious setback. LOUIS-PHILIPPE issued a decree dismissing General LAFAYETTE and disbanding the National Guard. The two republican batteries refused to be disarmed, and nineteen artillerymen, who were considered the ringleaders, were arrested. The men involved were: Captains CAVAIGNAC and GUINARD, and Privates TRÉLAT, SAMBUC, ANDRY, FRANCFORT, PENARD, ROUHIER, LENIBLE, PÉCHEUX, D'HERBENVILLE, CHAPARRE, GOURDIN, GUILLEY, CHANVIN, LEBASTARD, POINTIS, DANTON, the grandson of the famous GEORGES-JACQUES DANTON, and the GARDNIER brothers.

GALOIS, however, was not going to despair, and, the following year saw him fully active from the start. On 2 January 1831, the *Gazette des Ecoles* published a letter on the teaching of mathematics in Parisian institutions of secondary and higher education, beginning with the methods used to recruit teachers:

> Firstly, in science, opinions are unimportant.
> I want to know whether a particular individual is a good or bad teacher, and I do not worry about his opinions on questions unrelated to his scientific field. It was therefore frustrating and scandalous to see how, during the Restoration, teaching posts were available to those who professed the most acceptable royalist and religious views. The situation has not improved. Mediocrity ... is still rewarded.

Subsequently, though having himself had an excellent teacher in RICHARD, he criticized the way in which mathematics was taught, and the points he raised, though 160 years have passed, are still surprisingly relevant today:

> Let us begin with high schools, from which pupils studying mathematics mostly aim to go on to the *Ecole Polytechnique*. What is done to assist them in this purpose? Is any attempt made to enable them to understand the

Fig. 18: LOUIS PHILIPPE and his family in the gardens of the Château de Neuilly

true spirit of science, through explanation of the simplest methods? Is any effort made to make reasoning a second memory for them? Is there not, instead, some similarity with the way they learn French and Latin? [...] For how much longer will these wretched youngsters be forced to do nothing but listen and repeat all day long? When will time be spared to reflect on this enormous mass of knowledge, so as to establish a cohesive pattern in these numerous isolated propositions, in these unrelated calculations? Would it not be of use to the students if the simplest and most productive methods, calculations and forms of reasoning were expected from them? No. Truncated theories, overloaded with pointless observations, are taught in the minutest detail, while the simplest, most brilliant algebraic propositions are neglected. Actually ... great efforts are made to demonstrate ever longer, occasionally false calculations and arguments, and even obvious deductions.

Publishers are also attacked:

What is the origin of this negative situation? [...] Booksellers want thick tomes. The more there is in the books written by examiners, the more profit they will make from sales. This is why, every year, we see these huge new compilations, containing work taken from the great masters next to schoolboy essays.

Memories of his failing the entrance examinations to the *Ecole Polytechnique* led him to observations on the way examinations were conducted:

> Why do examiners ask such confusing questions? They seem to be afraid of being understood by candidates. What is the origin of this deplorable habit of compiling artificially difficult questions?

There is hardly a student who would not agree with GALOIS when he states:

> Thus it would be right to say that a new academic subject has recently been established, whose popularity is growing. It consists of knowledge of the scientific likes and dislikes, of the various eccentricities and temperament of individual examiners.

Nevertheless, GALOIS was to have no further dealings with examinations and examiners, since, on 4 January, two days after his letter had been published, a decree issued by the Royal State Education Board, confirmed his expulsion from the *Ecole Normale*:

> Following the reports written by M. Cousin, member of the Board, and by M. Guigniault, director of the *Ecole Normale*, regarding the provisional expulsion of Galois, and supporting evidence,
> it is decreed that:
> Galois leave the Ecole Normale immediately.
> Further measures will be taken regarding his future placement.

GALOIS had been receiving a grant at the *Ecole Normale*. Now he had no financial means, and his mother, who had difficulty in making ends meet with her job as a companion, was in no condition to keep him. The only way out was to give mathematics lessons. He made arrangements in a few days. CAILLOT, the bookseller, whose shop was in rue de la Sorbonne, meaning that he knew many students, decided to give him a hand, by allowing him to use a room next to the main room of the bookshop, and by promising to find him some pupils. GALOIS' aims were more ambitious, however. He was planning an advanced course, during which he would communicate his innovative research, most of which had not been published. Despite this, he was not afraid of others stealing his ideas. He was probably not particularly concerned, since he believed that science belonged to everybody. Thus, he placed the following advertisement in the *Gazette des Ecoles*:

> Evariste Galois, ex student of the *Ecole Normale*, will hold an algebra course, aimed at those students who wish to investigate the subject further, since this branch of mathematics is not exhaustively dealt with in state schools. The course will present some new theoretical aspects, none of which have yet been published or been the subject of public lectures. Here we shall only mention a new theory of imaginary numbers, the the-

ory of equations solvable by radicals, number theory and that of elliptical functions treated as pure algebra.

The lessons will be held on Thursdays at 13.15 at the Caillot bookshop, 5 rue de la Sorbonne. The course will begin on Thursday 13 January.

Who went to GALOIS' first lesson? The CHEVALIER brothers, certainly, and many republican friends. There seem to have been about forty people present at this first lesson, but were there any mathematicians among them? There is no way of knowing. Nevertheless, POISSON had asked GALOIS for a further copy of his research, so as to be able to present it to the *Académie des Sciences*, for the third time. GALOIS was uncertain about what to do, and waited a couple of days. Then, on 16 January, he wrote a new introduction to his *mémoire*, and took it to the secretary's office. It was presented during the following day's session, and LACROIX and POISSON himself were entrusted with examining it, and making a report. GALOIS was asked to wait.

His course was not a great success. It was practically incomprehensible to his friends, who were present mostly out of solidarity, and even those with some grounding in mathematics found his approach too unconventional to be able to follow it. Thus the number of listeners steadily decreased, until GALOIS found himself talking to an empty room. This meant that he was forced to restrict himself to what he considered trivial, private lessons for bored schoolboys and students.

It was probably at this time that he began to go to the lectures organized by the Academy, thus becoming better known in mathematical circles. His contributions to discussion were no doubt very acute and relevant, but at the same time aggressively critical, and certainly did not comply with the rules of academic etiquette. In 1831, SOPHIE GERMAIN, one of the few competent woman mathematicians of the period, wrote to GUGLIELMO LIBRI:

> ... he has kept up his capacity for being rude, a taste of which he gave you, after your best lecture at the Academy.

In the meantime, as no news had arrived about the *mémoire*, GALOIS sent a letter of enquiry to the president:

> I dare to hope that M. Lacroix and M. Poisson will not be offended if I remind them of a *mémoire* dealing with the theory of equations, with which they were entrusted three months ago.
>
> The research making up this *mémoire* is part of a work I had submitted for the *Grand Prix de Mathématiques* last year, in which I provide, rules for recognizing, in all cases, whether an equation is solvable or not by radicals.

Since this problem, has long appeared to geometricians if not impossible, at least very difficult, the prize committee decided a priori that I could not have solved it, firstly because my name is Galois, and also because I am a student. I was informed that my *mémoire* had been lost. This should have been a lesson for me. Nevertheless, I partially rewrote it, and submitted it to you, on the advice of a fellow of the Academy.

Sir, my research has evidently suffered the same fate of that of those who attempt to square circles. Will the analogy go further? Sir, I should be grateful if you could relieve me of my apprehension, by inviting M. Lacroix and M. Poisson to state whether they have also lost my *mémoire*, or whether they intend to report on it to the Academy.

Trouble with the Law

Three of the nineteen artillerymen considered responsible for the refusal to hand over their arms, when the National Guard had been disbanded in December, were released a few days after their arrest, but, in April, sixteen of them appeared before the Assize Court of the Seine. The trial, which came to be known as the "trial of the nineteen," attracted a great deal of attention in Paris. A large crowd, mostly made up of workers and students, crowded the Paris Law Courts, milling around the entrance to the court where the trial was scheduled to take place. The accused, who had been greeted with cheers as they came in, instead of trying to defend themselves, began an attack on the government. The speeches made in court were the best occasion yet for republican propaganda. Betrayal of the principles of the Revolution was denounced, as was the poverty of ordinary people.

CAVAIGNAC's speech actually consisted of a description of the party programme. The defence lawyers, who were all republicans, persuaded the jury that this refusal to hand over arms was not to be taken as a plot to overthrow the monarchy, and replace it with a republic, but only as a gesture of loyalty to the military body, in the ranks of which the nineteen had fought so many battles. Even old General LAFAYETTE came to testify for the accused, all of whom he knew personally. On arrival, he was warmly, and also respectfully, greeted by them. The trial ended on 16 April, with a verdict of not guilty. The nineteen were acclaimed by the crowd as true heroes. PÉCHEUX D'HERBENVILLE was given an especially warm reception, and his carriage pulled through the streets of Paris, "among shouts and wild applause."

To celebrate this victory, the members of the *Société des Amis du Peuple* organized a collection to pay for a banquet in honour of the artillerymen. The required sum was soon put together, and the banquet was scheduled for 9 May, in a room giving onto the garden, at the *Aux Vendanges de Bourgogne* Restaurant in the faubourg du Temple, in the Belleville district. There were 200 names on the guest list, including those of RASPAIL, MARRAST, ALEXANDRE DUMAS, ETIENNE ARAGO, and GALOIS, of course. As DUMAS wrote in *Mes Mémoires*: "it would have been difficult to find two hundred guests in the whole of Paris who were more hostile to the government." Many of them were provocatively wearing their National Guard uniforms. The atmosphere soon became very lively, "the champagne corks exploded like a serious exchange of artillery fire, and excitement was rife."

MARRAST had prepared the texts for the toasts in advance, and they had been approved by HUBERT and RASPAIL, with the purpose of preventing police provocation, but others were quickly added, especially by the younger guests. DUMAS was loudly encouraged to speak, which he did, but with caution:

"I drink to Art! May both pen and brush contribute as much as rifle and sword to the social renewal to which we have dedicated our lives, and for which we are ready to die!". This toast was applauded, while the words of ETIENNE ARAGO were greeted with shouts of enthusiasm: "I drink to the Sun of 1831. May it be as warm as that of 1830, but not blind us as the other did!". From every corner of the room the answer came: "Go back further!". Other toasts were proposed, to the Revolution of 1793, the Jacobin extremists of the "Mountain," and to ROBESPIERRE.

GALOIS, who was sitting at the far end of the long table, suddenly got up, raised a glass of wine in one hand, and brandished a jackknife in the other, shouting: "To Louis-Philippe!". Most of the guests followed his example, also making threatening gestures: "To Louis-Philippe!". This was followed by a certain amount of confusion. Some of the guests, frightened by the possible repercussions of those words, hurriedly left the room. DUMAS and an actor from the Royal Theatre, who was sitting next to him, and who did not wish to get involved, escaped through the window. The banquet came to an end in chaos.

The following day, the police arrived at the apartment GALOIS was sharing with his mother, and arrested him, on a charge of incitement to an attempt on the life and person of the King of the French.

Fig. 19: ALEXANDRE DUMAS

Fig. 20: Portrait of FRANÇOIS-VINCENT RASPAIL (1794–1878) as an old man. He can be considered a precursor of cellular theory and pathology, as well as one of the founders of cytochemistry. He made a special contribution to the spreading of basic notions of hygiene and medicine.

He was imprisoned at Sainte-Pélagie. The day after, he wrote to his loyal friend, AUGUSTE CHEVALIER:

> Under lock and key!! I am responsible for that gesture, but don't chide me, because the wine had made me lose my head.

GALOIS was not the only one to be arrested that day. The president of the Friends of the People could not be considered unconnected with the event. It was inconceivable that the King could be threatened so easily. The French people needed to be shown that any gesture that undermined the good name of the Crown would not go unpunished. A different charge, however, was found for RASPAIL. A long forgotten letter that he had sent to *La Tribune*, the previous February, was declared to be offensive to the King and the National Guard. RASPAIL was tried quickly, given a prison sentence of eight months and fined 800 francs.

GALOIS' trial, on the other hand, began on Wednesday 15 June, before the Assize Court of the Seine. He was defended by DUPONT, a republican lawyer and member of the *Société des Amis du Peuple*. The trial opened with questions to the accused from the presiding judge, NANDIN. GALOIS was asked to specify the occasion on which he had threatened the King's life, whereabouts in the room he was sitting, and what toasts had been proposed before his. To the question: "Did you not take out a knife from under your jacket and say the words 'To Louis-Philippe!'," he answered: "This is what happened: I had a knife which I had used to cut meat during the meal. I brandished it saying 'To Louis-Philippe, if he betrays us!' These words were only heard by the people next to me, considering the whistling that had begun after the beginning of my utterance, because people thought I was proposing a toast to the good health of Louis-Philippe."

The judge was astonished by what GALOIS said, and asked him why he was afraid the King would become a traitor. "Everything encourages us to adopt this position," he answered, "... it is reasonable to believe that Louis Philippe could betray the Nation. He has not given us enough guarantees ... all the King's actions, though not yet showing his bad faith, can lead us to doubt his good faith. One example is the background of intrigue to his accession to the throne."

At this stage, DUPONT pointed out that the judge's questions were touching on problems that would force the defence to provide explanations he would prefer to avoid. MILLER, the public prosecutor, agreed with DUPONT. Discussion then passed on to the knife, which GALOIS had purchased from HENRY, the knife maker, on 6 May. It was a jackknife which he had wanted for some time, but had only been able to afford to buy three days before the banquet, at the price of fourteen francs. Madame HENRY, the knife maker's wife, confirmed his story.

Then the witnesses for the prosecution were called. M. PETIT, representative of the officials of the court of primary jurisdiction, stated that, on 9 May, he was in a room next to the one where the republicans' banquet was being held. He had heard the toasts and shouts, but was unable to report GALOIS' actual words. M. DELAIR, attorney of the Royal Law Courts, who had been one of the two hundred guests, testified to the fact that he had seen the accused stand up at the other end of the table, holding something shiny, which could well have been a knife blade. He had not heard the words "if he betrays us" after the toast, but could not say that they had not been uttered.

Meanwhile, noise outside the court room interrupted the proceedings. The Assize Court officials had not allowed the large number of journalists in the building access to the court room. In answer to their violent protests, the presiding judge gave those journalists with proof of their status permission to follow the proceedings.

It was now the turn of another witness, M. DENIS, head waiter at the restaurant where the banquet had been held, who stated that he had heard nothing of interest, except laughter and singing. M. DURANDON and M. DESESQUELLE, both waiters, said that they had heard the words "republic" and "revolution" several times, but were unable to be more precise, since they had been busy putting away the silver. ROUX, the wine waiter, was also called, and he claimed to have heard a toast to the "Republic of 1831."

COUET, PERON and CRETON, all three law court officials, agreed with PETIT's testimony. M. GUÉRET, a butcher, said he had heard the shout: "Long Live the Republic!." He added that they were a rowdy crowd, who had even dared to smoke in the large *salon*, something that had never happened before. GUÉRET even stated that he had heard someone shout: "Death to Louis-Philippe, and the guillotine for all his family!". Though he had not been invited to speak, GALOIS attacked GUÉRET, who, he claimed, was the one who had threatened some of the guests, especially M. EUGÈNE PLANIOL, a man of letters. PLANIOL was in court and willing to make a statement to this effect.

When called to testify, M. GUSTAVE DROUINEAU, who was wearing a decoration awarded during the July uprising, refused to take the oath. He said that he had no intention of telling other people about what had happened at a private banquet. He added that, though not wishing to challenge justice, he believed he had the right to refuse to answer the questions. He was fined 100 francs, on the basis of article 80 of the penal code.

The accused had asked for LECOMTE, SOUILLARD, BILLARD, AUDOUIN and CUPER as witnesses for the defence. They all declared that the first words of GALOIS' toast "To Louis-Philippe!" had caused something of a commotion among the guests, which prevented the follow up, "if he betrays his oaths," from being heard. However, as they were sitting next to him, they had heard everything. BILLARD, a pharmacy student, added: "We were talking about what would have happened, if somebody had proposed a toast to Louis-Philippe. M. Galois said he would propose the toast 'To Louis-Philippe, if he betrays his oaths!'. We pointed out to him that we did not know the oaths in question, and, while we were talking about when the people would recover their rights, he proposed the toast, as he had said he would. Some people whistled, not having heard the last words, but, as soon as they were understood, they were greeted with enthusiastic applause."

The last to be heard were HUBERT and RASPAIL. The latter's presence caused considerable curiosity among those present, since he had recently refused a first class decoration from the King. Both claimed that the words of the toast had not been intended to provoke a new uprising.

When the public prosecutor began his concluding speech, he raised the question of whether the republicans' meetings were public of private. The banquet had been theoretically private, but witnesses had shown that most of what had been said in the room could be heard outside, thus making the meeting a public one. Precedents from the High Court of Appeal established that restaurants and hotels were public places. Therefore, the accused had committed the crime of incitement to an attempt on the person and life of the King, at a public meeting.

GALOIS was then allowed to speak: "I shall respond to some of the public prosecutor's errors. My answers in the preliminary hearings were objected to, as was the omission of the words 'if he betrays his oaths.' I must admit that I preferred to comply with the wishes of the judge at the preliminary hearings, rather than run the risk of staying in prison for three or four months. I confess my behaviour was rather sly. You can surely imagine the police inspector's joy, when he thought he had unmasked a conspirator. He thought his name was made! He must be rather disappointed by now. I cannot let what the public prosecutor said about it being impossible for the King to be a traitor go unanswered. Nobody is foolish enough to believe now that a king is perfect, especially since when judges, who under Charles X, persecuted us, because we said that the King could neglect his duties, have themselves now sworn allegiance to another man, who had been placed on the throne, as the result of his predecessor's stupid behaviour." He added that he was one of those who had been walking the streets of Paris armed, and who would like to have been with his friends at the previous Saturday's hearing.

"Men of the Restoration," he went on, "observe the results of your actions. You promised that there would be no more uprisings, but they are still taking place! Charles X was far cleverer than you! You are children! You put our heads on the block, but have not had the courage to bring down the axe. We are children too, but children who go ahead, brimming with strength and courage. Our republicans will never know corruption!"

In GALOIS' own interest, the presiding judge wisely interrupted his speech. In his concluding remarks, the defence lawyer restricted himself to the question of whether the meeting had been public or not, insisting on the fact that whatever had been said was in private.

The jury was out for half an hour, at the end of which the answer to the judge's question arrived. The accused was not guilty. GALOIS was silent for a while, betraying no emotion. Then he calmly got up from his place, walked up to the table where all the objects admitted as evidence had been placed, picked up his knife, folded the blade back into place, put it in his pocket, and, without a word, left the court room.

The hearing he had mentioned in his speech had taken place on Saturday 11 June, as part of the trial of some of GALOIS' friends. MALOT and LEBON, both medical students, and BOUDEL, GRIVEL and MATHÉ, all law students, who had been decorated for their part in the July uprising, were the accused. The trial had investigated the events of Friday 11 March, which are particularly revealing as regards the usual activities of the Friends of the People. At 1 p.m., a group of about thirty young people had gathered in the square in front of the Panthéon. At first, the city police had dispersed them, but they soon regrouped, with the purpose of entering the building. When they discovered that all the doors were locked, they began to pick up stones and throw them at the main door, until the locks gave way. A dozen of them entered, and took down a number of tricolour flags, hanging at half mast by the funerary monument of BENJAMIN CONSTANT, and shared them out, leaving only one in place. On leaving the church, they walked along rue de la Montagne-Sainte-Geneviève, passing in front of the *Ecole Polytechnique*.

One of the most excitable young men pointed at it and shouted: "They won't come this time! They're too scared!". The group, which had now passed the two hundred mark, moved on into faubourg Saint-Antoine, with shouts of: "Long Live Freedom! Long Live the Republic!". There followed a violent clash with the police, leaving several injured on both sides. The five students were arrested, charged with resisting arrest, and unruly behaviour. MATHÉ was also charged with possession of illegal weapons. DUPONT defended the five youngsters, who were each given two month prison sentences.

Naturally, nothing had been said, during his trial, about GALOIS' exceptional mathematical talents. His defence counsel, did not even mention his problems with the Academy and the *Ecole Normale*. This aspect had been taken care of by the Saint-Simonian newspaper *Le Globe*, on 15 June, the very day the trial began. The idea had originated with the CHEVALIER brothers, one of whom may have been the author of the article:

> Young Galois will appear before the Assize Court today, charged with saying "To Louis-Philippe!", while brandishing a knife, during a banquet at the *Vendanges de Bourgogne*, on 9 May. Nothing could be further from our intentions than support for such a gesture. Our religion is one of peace, reconciliation, and order, and we stand back in horror in the face of any idea of violence and blood [...]. In all sincerity, we declare that even if a threat with a knife were subject to countless conditions, even if it were made in the most secret place, where the arm of the law does not reach, it would still bear a barbarous hue.

It was undoubtedly the non violent aspect of Saint-Simonianism that alienated GALOIS, who preferred the, occasionally bizarre, revolutionary approach of the *Société des Amis du Peuple*. The article continued:

> However, there are attenuating circumstances in young Galois' favour. It is our duty to communicate them in the interest of this unfortunate young man. [...]
>
> Though M. Galois is not yet twenty, he has provided ample, irrefutable evidence of great scientific ability. Nevertheless, despite all his efforts, he has been met with nothing but indifference and scorn for his talent. Considering himself a victim of the social order, he has become embittered, discouraged and frustrated. He felt the germs of a brilliant future within him, but, cast into the midst of a selfish society, with neither protectors nor friends, he nurtured violent hatred of a régime, under which the accident of birth hides so many promising faculties, while, on the other hand, the same accidental circumstances promote so many nonentities [...].
>
> Galois' high intellectual calibre is a constant factor. He discovered the properties of elliptical functions, at the same time as Abel, the northern

scientist whose merits were ignored by the *Institut de France*, until after his death in abject poverty.

GALOIS must have often complained to his friends, the CHEVALIER brothers, about still not having received a judgment on the *mémoire* on the solvability of algebraic equations by radicals that he had submitted to the Academy, for the third time, with the encouragement of POISSON. The mathematical establishment was again showing its total lack of interest in the ideas that he considered so new and important. Why not make such negligence public, through the pages of *Le Globe*, and thus force the distinguished fellows of the Academy to finally take notice of GALOIS' work?

> Before March of last year, M. Galois sent a *mémoire* to the secretariat of the *Institut de France* on the solvability of numerical equations. This *mémoire* was his entry for the *Grand Prix de Mathématiques*. It was considered unworthy, since it overcame certain difficulties, to which even Lagrange had not been able to find solutions. M. Cauchy had showered praise on the author's treatment of this problem; but what did that matter? The *mémoire* was lost, and the prize awarded, without the young scholar being able to take part in the competition. In reply to a letter to the Academy from young Galois, complaining about the negligent treatment of his work, all that M. Cuvier could write was: "The matter is very simple. The *mémoire* was lost on the death of M. Fourier, who had been entrusted with the task of examining it". The *mémoire* was rewritten and submitted to the Academy again. M. Poisson, who is to examine it, has not yet done his duty, the result being that its wretched author has been waiting for a kind word from the Academy for more than five months.

Making the matter public forced LACROIX and POISSON to act. Their report was presented at the 4 July session. Their judgment, however, was negative. Despite not having understood what GALOIS had written, their opinion was that his work was wrong. But it is *their* report that contains an error, as any reader familiar with mathematics will notice, showing their total inability to accept GALOIS' new, revolutionary methods of investigation:

> ... it should be noted that [the theorem] does not contain, as the title would have the reader believe, the condition of solvability of equations by radicals ... This condition, if it exists, should have an external character, that can be tested by examining the coefficients of a given equation, or, at most, by solving other equations of a lesser degree than that proposed. We made all possible efforts to understand M. Galois' evidence. His thesis is neither clear enough, nor sufficiently developed to enable us to judge its rigour.
>
> Neither are we able to provide a clear idea of this work. For this reason, we return your manuscript in the hope that you will find M. Poisson's observations of use for your future research.

Fig. 21: SIMÉON-DENIS POISSON (1781–1840). His research particularly covered celestial mechanics, electrostatics, magnetism and probability theory.

The letter was signed by FRANÇOIS ARAGO, the secretary of the Academy. He was the same ARAGO who had tried to mediate between the people and General MARMONT, on 28 July.

GALOIS' only consolation was that he had got his manuscript back. His problems with the law had strained relations with his mother. Furthermore, the total lack of understanding of the ideas that were to constitute the foundation of modern algebra, and that were all the more astonishing, in that they had been formed in the mind of a seventeen-year-old, had discouraged him to the extent of making him want to live alone, thus enabling him, when he thought he needed it, to seek refuge in physical isolation. For this reason, he decided to rent a room at 16 rue des Bernadins.

In Paris, plans for the celebrations of 14 July 1831 were under way. The republicans had decided to organize a patriotic demonstration in place de la Bastille, during which a tree of liberty would be planted. A poster was prepared inviting workers, students and young shopkeepers, all "July men," to a gathering in place du Châtelet, at noon on Thursday 14 July. Before 1 p.m., a march along the Seine embankments, rue Saint-Martin and the boulevards, led by a band playing patriotic songs, would escort the tree as far as place de la Bastille. Present and ex members of the National Guard were asked to take part in uniform. Worried about the consequences for public order of the planned demonstration, VIVIEN, the prefect of police, had all the copies of the poster at MIE the printer's seized, and decided to arrest those republicans who were considered most subversive. Naturally, GALOIS was one of them. The police broke into their houses during the night of 13 July, but most of them had been warned, and were not at home. GALOIS did not spend that night in rue des Bernadins.

Nevertheless, the following day, obeying party orders, he set off, with his friend VINCENT DUCHÂTELET, at the head of group of six hundred demonstrators. At about 12.30, the group, from rue de Thionville, the present rue Dauphine, began to cross the Pont Neuf. Half way across the bridge, they were stopped by the police, who dispersed them in a few minutes. GALOIS and DUCHÂTELET were arrested. The former was wearing his old National Guard uniform, and carrying several pistols and his usual knife, apart from his regulation carbine. There was no problem in finding a charge against him. The same morning Generals DUBOURG and DUFOUR were also arrested. In the Champs-Elysées, the republicans had been attacked by a rival group of demonstrators, violent scuffles continuing all day long.

GALOIS and DUCHÂTELET were locked up in the cells at the police station in rue de Thionville, where the latter played a prank that only made their position worse. He drew a pear on his cell wall. In the political cartoons of the time a pear represented the King's head, and DUCHÂTELET added a guillotine with the words

To Louis-Philippe!

Fig. 22: View of the Pont Neuf

"Philippe will take his head to your altar, Freedom!" written below it. In the evening, they were both transferred to the Sainte-Pélagie prison, where GALOIS was given the number 15348. RASPAIL, who had been arrested on the charge of conspiring against the King, as a member of the Friends of the People and also "with pen and ink," had been in the same prison since May.

After three months of preventive detention, GALOIS was to pay for his being found not guilty in June. He was transferred to the Conciergerie for the duration of his trial, which began on 23 October. The trial was deliberately not held before the Assize Court, since a verdict of guilty was essential for the authorities. He was charged with possession of illegal arms, and wearing a uniform to which he was not entitled. Neither GALOIS nor DUCHÂTELET had been included in the reorganization of the National Guard, which had been "set up again to defend the constitutional monarchy, the *Charte*, and the rights listed in it, to ensure respect for the law, to re-

Fig. 23: LOUIS-PHILIPPE "King of the French"

establish and maintain order and public harmony and support troops under arms." Actually the new National Guard was made up of those citizens who paid direct taxes, and who were able to pay for their own equipment. It was now an exclusively middle class institution.

DUCHÂTELET was sentenced to three months, and GALOIS, with whom the judges wanted to be particularly strict, to nine. GALOIS appealed, but his sentenced was confirmed on 3 December. His stay in the Sainte-Pélagie prison would last until April 1832.

The Trial of the Fifteen

Following the disorders in July, the police decided to punish those leading members of the *Société des Amis du Peuple* who had not yet ended up in Sainte-Pélagie, and lengthen the sentences of those who were already there. Fifteen members were tried before the Assize Court, charged with incitement to hatred of and scorn for the government, and to its overthrow by means of hostile propaganda. The accused were: RASPAIL, BLANQUI, THOURET, HUBERT, TRÉLAT, BONNIAS, RILHEUX, PLAIGNOL, JOUCHAULT, DELAUNAY, BARBIER, PRÉVOST, RIVAIL, CHOIGNEAU, and GERVAIS. The last was also charged with resistance to a police officer.

Many years later, RASPAIL wrote in his autobiography:

> All of us, as leading members of the Society of Friends of the People, at our own risk, had written dozens of articles dealing with our programme: universal suffrage, open access to all public office, freedom of the press, graduated taxation. It was for this programme that we were brought before the Assize Court. It was because we had attacked Louis-Philippe for having betrayed his pledges, for the insult to the Revolution, for the people who had been deceived, and for forgetting the barricades.
>
> The authorities loudly proclaimed that the Society of Friends of the People was a secret organization, then whispered in every ear that a secret society conspires, by its very nature, against order, and law abiding citizens' safety. We were about to show the republicans' strength.

The hearings took place on 10, 11 and 12 January 1832, attracting considerable attention, all over France. Detailed reports of the speeches of both prosecution and defence appeared in *Le Temps*, *Le National*, *Le Mouvement*, and *La Tribune*. The presiding judge was JACQUINOT-GODARD, magistrate of the Royal Law Court in Paris, and the assistant public prosecutor DELAPALME, formerly a faithful subject of CHARLES X, who had managed to keep his post, thanks to his declarations of loyalty to LOUIS-PHILIPPE. Naturally, DUPONT was one of the many defence lawyers. More than fifty witnesses had been called by the latter alone. Among them were GODEFROY CAVAIGNAC, BASTIDE and GALOIS, who, as a prisoner, was kept in an adjacent room under armed guard.

CAVAIGNAC explained that the Society's publications were collective, and that they had never conspired:

> In 1825 there was a town based secret society, originating from the *Carbonari*. One of its members was M. Barthe, the present Minister of Justice.

GALOIS was not allowed to speak for long, and all he could do was confirm CAVAIGNAC's statement. RASPAIL, on the other hand, made a long, detailed speech, which was not defensive, as would have been expected, but an emotional, resounding appeal on behalf of the people's needs, ending on a harsh note:

> The traitor bearing the title of King should be buried alive under the ruins of the Tuileries. This is all a citizen who demands fourteen million for living expenses of an impoverished France deserves.

BLANQUI was no less provocative. When asked his profession by the presiding judge, he replied "Proletarian." When the objection was made that this was not a profession, he added: "What do you mean, it is not a profession! It is the profession of 30 million Frenchmen, who live by their labour, and have no political rights."

At the end of the trial, after three hours' deliberation, the jury declared that the writings did in fact contain the crimes identified in the preliminary hearings. However, since they were anonymous, the accused were not guilty. RASPAIL and BLANQUI were sentenced, nevertheless, because of their public statements, the first to fifteen months, and the second to a year.

Thus BLANQUI joined RASPAIL and GALOIS in the Sainte-Pélagie prison.

Prison

The Sainte-Pélagie prison was situated in the Jardin des Plantes district, on the southern edge of Paris. The park, after which the district is named, contained botanical gardens and a zoo, which had been set up, when the animals from the royal menagerie in Versailles were moved there during the Revolution. Under CHARLES X, the Jardin des Plantes had become very fashionable. One of its attractions was a giraffe, donated to the King of France by MEHEMET ALI, Pasha of Egypt. This exotic animal was the favourite topic of conversation in Paris drawing rooms for several months, and had inspired dress designs and even hair styles.

The long lines of prisoners marching to Sainte-Pélagie were certainly not a pretty sight, and this drove away respectable citizens, who sought refuge in the Tuileries and Luxembourg gardens for their walks. The district was now just another ordinary suburb. Before 1789, the building housing the prison had been an institution for redeemed prostitutes, only taking on its new status in 1792. It covered the area enclosed by rue du Puits-de-l'Ermite, rue du Batton, rue Copeau and rue de la Clef. Gates gave onto the last of these streets, and the one parallel to it on the other side. Several floors of cells and dormitories surrounded and separated three large yards, in which slender acacia trees grew. The whole building was enclosed by a high wall, topped by patrolled battlements. There were six ground floor windows on the façade, and only narrow slits higher up, which looked like holes for scaffolding, that some absent minded builder had forgotten to fill in. The cells surrounding each yard were assigned to a particular category of prisoner. The political prisoners, mostly republicans, were housed in the first section, which looked onto rue du Puits-de-l'Ermite. Thieves and swindlers were in the middle section, while debtors were in the last, which was entered from rue de la Clef. There were many abandoned children, and tramps, who had been arrested when caught begging or stealing, in the middle section.

The political prisoners were treated differently, according to their financial circumstances. The richer among them could pay for a single cell, and have their meals brought in from a restaurant. At a lower price, a bed was available in a cell for seven or eight. The poorest were crowded into free sixty bed dormitories. GALOIS, in all probability, belonged to this last category. Journalists who had written articles hostile to the régime were also considered political prisoners, though, as a small élite of rather arrogant intellectuals, who were proud of suffering for their ideas, they were treated with special respect. CHATEAUBRIAND, for example, who had been sentenced for a number of articles he had written in *Le Conservateur*, of which he was both founder and editor, spent many tedious days in prison, being forced to listen to the prefect of police GISQUET's wife, pounding out waltzes on the piano, in the drawing room of their house next to the prison buildings. Of course, life for the other inmates was not quite so comfortable.

GALOIS spent much of his time in prison, walking up and down the yard, in meditation. A series of isolated sentences, belonging to a more cohesive context, which were very probably written at this time, have been found among his papers:

> As long as a man says: I am science, he must have a name to counter those he opposes. If not, his ambition will be seen as envy.
>
> A man who has an idea can choose between a very high, lifelong reputation as a learned man, or establish a school, be silent, and ensure the future survival of a great name. The first case occurs, if he puts his idea

into practice without enlarging on it, the second, if he publishes it. There is a third middle way: to publish and practise. Then one is ridiculous.

RASPAIL also spent most of his prison sentence studying and thinking. It was at this time that he wrote more than fifty letters, which were published, eight years later, and entitled *Lettres sur les prisons de Paris*. In many of them he mentions GALOIS, of whom he was very fond, and with whom he sympathized, and they show the deep admiration he felt for this ardent young prisoner:

> This slender, dignified child, whose brow is already creased, after only three years' study, with more than sixty years of the most profound meditation; in the name of science and virtue, let him live! In two years' time he will be Evariste Galois, the scientist! But the police do not want scientists of this calibre and temperament to exist.

For GALOIS, the best part of the day was the republicans' evening ceremony. They gathered around a tricolour flag, and sang the *Marseillaise* and other patriotic songs. After the singing, they marched past the flag, each of them kissing it. Occasionally amateur theatricals followed. These always consisted of allegorical scenes, written by the prisoners themselves, recalling the July Revolution. There was no scenery, and the only stage prop available was a coffin, which was used to carry the corpse of the republic murdered by LOUIS-PHILIPPE. Even though these ceremonies contributed to solidarity among republicans, young, thoughtful GALOIS, who kept very much to himself during the day, was continually made fun of by his fellow prisoners. They nicknamed him ZANETTO, and often forced him to drink spirits, daring him to drink down the whole bottle. Though disgusted by what he had to drink, GALOIS was unable to resist. He would get very drunk, and then faint. It was his friend RASPAIL, as the latter describes in one of his letters, who helped him to bed:

> He was wandering round the prison yard, one day, deep in thought, in a kind of day dream. His bleak looks were those of a man who was only physically present on earth, and is only kept alive by his thoughts.
>
> Our bully boys shouted out: "Hey, you may be only twenty, but you're an old man! You can't take your drink, can you? Drinking scares you, doesn't it?" Then he marched straight up to the danger, downed a whole bottle, all at once, and threw it at his tormentor. He would have been justified, if he had killed him there and then!

RASPAIL is the only source for two significant events during GALOIS' imprisonment. On 27 July, when he had only been at Sainte-Pélagie for thirteen days, a high mass was celebrated in memory of

Fig. 24: FRANÇOIS-VINCENT RASPAIL at the time of his imprisonment in Ste-Pélagie

the victims of the previous year's uprising. A revolt by the political prisoners was expected for 28 and 29, and the prison warders were on the alert. Nothing actually happened. On the evening of 29, just after the prisoners had returned to their dormitories, one of them, who occupied a bed a short distance from GALOIS', was wounded by a bullet fired from a house opposite the prison. Who was the bullet meant for? Were the police trying to rid themselves of GALOIS, once and for all? Was he considered so dangerous? No answers are forthcoming to these questions. The prison governor was accused of being behind the episode, and the expected revolt did take place. GALOIS was transferred to a punishment cell.

RASPAIL then provides a rather unclear account of a suicide attempt by GALOIS. He describes a scene, during which young "ZANETTO," perhaps under the influence of drink, shouted the following, rather puzzling words: "You are supposed to be my friends, but you scorn me! You are right, but someone like me, who is responsible for such a crime [what crime?], must kill himself." RASPAIL continues: "He would have done just that, if we had not grabbed him, since he was holding a weapon." It would be pure speculation to take these matters further.

While in prison, GALOIS naturally continued his mathematical research, and, in October 1831, wrote a preface for his *mémoires*. It makes fascinating reading, needing no commentary, but has been kept out of print for a long time by the scientific establishment, since, as LEOPOLD INFELD observed, it is a violent accusation against a scientific hierarchy which placed pride before humility and arrogance before goodness.

When, in 1906, JULES TANNERY edited those of GALOIS' writings that had not yet been printed, he deliberately left out this preface, since, in his view, its author must have been either drunk, or feverish when he wrote it.

> Firstly, the second page of this work is not crowded with surnames, first names, titles, rank and eulogies of some miserly prince, whose purse strings would be loosened when incense was being swung, only to be tightened up when the censer was empty. Neither is respectful homage to be seen, in letters three times the size of the title, to some high ranking scientist, whose protection is indispensable (I nearly wrote, inevitable) for any twenty year old aspiring writer. I do not acknowledge anybody's advice or encouragement as being responsible for the good qualities of my work. I would be a liar, if I did so. If I had a message for the great men of the world, or the great men of science (and, in the present situation, the distinction is practically non-existent), I swear it would not be one of thanks.

It is the men of science who are responsible for the late appearance of the first of the two *mémoires*, and the other category, for the fact that all this was written in prison. During my detention, which is wrongly thought of as an occasion for reflection, I have often felt astonished at how I neglected to keep the mouths of my stupid "Zoili"[1] shut. I believe I am right to use the term "Zoilus," in all modesty, so low is my opinion of my opponents. Saying how and why I am being kept in prison is not part of my argument here, but I must mention how my manuscripts got lost in the files of the fellows of the *Institut de France*, though I fail to understand such negligence by men who have Abel's death on their conscience; not that I wish to be compared with that eminent geometrician. Suffice it to say that my *mémoire* on the theory of equations was submitted to the *Académie des Sciences* in February 1830, that preliminary extracts had been sent in 1829, that I was given no report on it, and that I never managed to get my manuscripts back. There are other similar anecdotes, but it would be silly of me to report them, since nothing happened to me, except for my manuscripts being lost. My bad character, a fortunate traveller, saved me from the wolves' jaws. I have already said enough to enable the reader to understand why, despite my willingness, it was impossible for me to adorn, or mar, as the case may be, my work, with a dedication. Secondly, the two *mémoires* are short, not at all doing justice to their titles. There is also as much French as algebra, to the extent that, when the printer saw the manuscripts, he thought they were an introduction. This was unpardonable on my part. It would have been so easy to refer back to a whole theory in its elements, with the pretext of presenting it in a way that was necessary for understanding the work, or, better, without the way of filling a branch of science with two or three new theorems, without saying which ones! It would also have been so easy to subsequently substitute all the letters of the alphabet in each equation, numbering them in order, so as to recognize to which combination of letters the following equations belong. This would have multiplied the number of equations indefinitely, if one recalls that, after the Latin alphabet, we have the Greek one, and, when this has been used up, we still have German Gothic letters, and nothing would stop us using Syriac, or even Chinese lettering! It would have been so easy to transform each sentence tenfold, making sure that each transformation was preceded by the solemn word "theorem," or by OUR ANALYSIS, achieve results that have been known since the time of good old Euclid, or, finally, place a fearful pack of special examples before and after each proposition! I was unable to choose one from so many ways! Thirdly, the first *mémoire* is not entirely unseen by an experienced teacher's eye. An extract sent in 1831 to the *Académie des Sciences* was examined by M. Poisson, who even admitted, during an official session, that he had not understood it at

---

1 ZOILUS was a 4th century B.C. grammarian from Amphipolis, well-known for his strictures against HOMER. The name is used, particularly in French and Italian, as a synonym for a carping critic.

all. This only proves to the eyes suffering from the self admiration of the author himself that M. Poisson did not want to, or could not understand, but will undoubtedly prove to the eyes of the general public that my work means nothing.

All this leads me to believe that the work I am presenting to the public, will be received in the scientific world, with a smile of sympathy, that the most generous will attribute my failure to lack of ability. For a period, I shall be compared with Wronski,[2] or those men who untiringly find a new solution to the squaring of a circle, every year. I shall have to bear, especially, the uncontrollable laughter of the examiners of candidates for admission to the *Ecole Polytechnique* (all of whom I am surprised not to see occupying places in the *Académie des Sciences*, since their names are not for posterity), who tend to have the monopoly of printing mathematics books, and will be irritated to learn that a young man they rejected twice has the presumption to write books that establish new ideas rather than mere textbooks.

All the above I have said to show that I am fully conscious of the fact that I am exposing myself to the mockery of the foolish. If, despite such small likelihood of being understood, I am publishing the end product of my sleepless nights, nonetheless, it is to date my research precisely, and to let those friends I made in the outside world, before I was buried under lock and key, know that I am still very much alive. I also hope that this research may fall into the hands of people who will not be prevented by stupid conceit from reading it, and who will be guided in the new direction that, in my view, analysis must follow in its highest branches. It should be pointed out here that I only mean pure analysis. If my assertions were to be transferred to more direct mathematical applications, they would become paradoxical.

Early on, long algebraic calculations were of little use for progress in mathematics. Very simple theorems had little to gain from being translated into the language of analysis. It is only since Euler that this briefer language has become indispensable for the new extension that this great geometrician gave to science. From Euler onwards, calculations have become more and more necessary, but more and more difficult, as they were applied to more advanced scientific subjects. Since the beginning of this century, algorithms had become so complicated, that progress was impeded by this means, without the elegance with which modern geometricians have endowed their research, and, by means of which, the mind quickly grasps, all at once, a large number of operations.

It is clear that this elegance, of which they are justly proud, has no other purpose.

On the basis of the well documented fact that the efforts of the most

---

2 HOENE WRONSKI (Poznan 1778–Paris 1853), in 1812, published a short article entitled *Résolution générale des équations de tous les dégrés*, obviously containing serious errors.

progressive geometricians are directed towards elegance, it can be certainly concluded that it is more and more necessary to cover several operations, at the same time, because the mind no longer has time to pause over details.

Now, I believe that the simplifications due to elegance in calculations (intellectual simplifications, of course, material ones being absent) are limited. I believe that the time will come when the algebraic transformations foreseen by analysts' speculations will no longer have the space or time for production; to such an extent that foreseeing them will have to be sufficient. I do not mean that there is nothing new for analysis, without this help, but I do believe that, one day, without this, all will be consumed.

In my view, the task of future geometricians is to stand squarely by these calculations, to group the operations, according to difficulty and not according to form. This is the direction I have begun to follow in this work.

My opinion here should not be confused with the conceit shown by some in avoiding all calculations, translating what is expressed very briefly by algebra into very long sentences, thus adding inappropriate linguistic complication to long operations. These people are a hundred years out of date.

There is nothing like that here. Here analysis of analysis is carried out. Here the highest calculations carried out up to now are considered special cases, which it was useful, even indispensable to examine, but which it would be tragically mistaken not to abandon for wider research. It should be time to carry out calculations required by this high analysis, and classified according to difficulty, but not specified as to their form, when the specificity of a question claims them.

My general thesis will be understandable only after my work, which is an application of it, has been carefully read; not that this theoretical viewpoint preceded application. However, after finishing my work, I asked myself what would make it so strange for most readers, and on reflection, I think I observed this inclination in my mind to avoid calculations in the subjects I dealt with, and, besides, I acknowledged an unsurmountable difficulty for anyone who wished to carry them out, generally, in the subjects I dealt with. It is to be expected that frequent difficulties, I was unable to overcome, would present themselves in treating such new subjects, and venturing along such a new path. For this reason the formula "I do not know" is often to be found in both these *mémoires*. This is especially the case with the second one, which is the most recent. The readers I mentioned at the beginning will have reason to complain. Undoubtedly, the most precious book by the most learned author, is the one in which he includes everything he does not know. It is also undoubtedly true that the greatest harm an author can do his readers is when he disguises difficulties.

When competition, i.e. selfishness, no longer dominates science, when scientists form research groups, instead of sending sealed packages to academies, even small, recent discoveries will be hurriedly published, and the words "I do not know any more" will be added.

GALOIS' sister NATHALIE-THÉODORE was a frequent visitor to Sainte-Pélagie, where she tried as hard as she could to alleviate her brother's suffering, with tasty food and pleasant conversation. In December 1831, however, she could not avoid noting in her diary:

> Five more months without fresh air! It is a very sad outlook, and I really fear that his health will greatly deteriorate; he is already so strange!
>
> He has no distracting thoughts, and has become gloomy, ageing before his time. His eyes are sunken, as if he were fifty!

Though "strange" and "gloomy," GALOIS was still the affectionate youngster he had always been. Even in prison, he did not forget the people who loved him, in whom he showed his customary interest. In January 1832, he wrote to CÉLESTE-MARIE GUINARD, his mother's sister:

> My dear aunt, I have been told you are ill and bedridden. I feel the need to let you know how sorry I am, and this feeling is all the more acute, in that I am deprived of the pleasure of seeing you, since I am confined to my room, and can visit nobody.
>
> You were kind enough to think of sending me presents. It is very pleasant to receive reminders of the living, in a tomb.
>
> I hope you will be in good health, when I leave prison.
>
> My first visit will be to you.

Though many leading members of the *Société des Amis du Peuple* were in prison, the republicans were unable to realize their plans for a new popular uprising. There had been many attempts, and the police had been very busy in the sixteen months following the July Revolution. Four prefects of police had come and gone: GIROD DE L'AIN, BAUDE, VIVIEN and GISQUET. However, in the spring of 1832, the republicans' activities had been checked in Paris, and all over France, by the psychological effects of a terrible cholera epidemic. The economic crisis and social conflicts were forgotten, in the wake of the fear of disease. Deaths in Paris were particularly numerous in the central districts, where the poorest inhabitants lived, in extremely bad hygienic conditions. Most middle-class families left the city for the isolation of their homes in the country.

It was also decided to transfer the youngest prisoners, and those in bad health, from Sainte-Pélagie, so as to restrict the danger of contagion. On 16 March 1832, GALOIS was transferred, as a prisoner on parole, to a clinic at 86 rue de l'Oursine, the present rue Broca. The clinic was named after FAULTRIER, its owner, and a doctor by the name of JEAN-LOUIS POTERIN-DUMOTEL, who lived with his family in the same street, worked there.

This transfer brought a great novelty into GALOIS' life. He met STÉPHANIE, POTERIN-DUMOTEL's daughter, and fell in love with her. Was she also in love with him? What sort of relationship actually existed between them? The only evidence consists of a copy of two letters from STÉPHANIE, that GALOIS must have torn up in a moment of rage. He regretted this later, and tried to reconstruct the content. The handwriting is GALOIS', but, at the end, the name Mademoiselle STÉPHANIE D. appears. There are some blank spaces between words. Despite these gaps, the two texts are mostly coherent.

> Please, let us put an end to this
> I do not have the spirit to keep up
> such a correspondence
> but I shall try to find enough to
> converse with you as I did before
> anything happened. Here
> the
> has   that
> owes you   than to
> me and not think things
> that may not exist and that
> will never exist.

It seems, then, that, initially, STÉPHANIE had encouraged GALOIS' love, or, at least, she appeared to him to be willing to accept it. What happened then? It was probably GALOIS' declaration of love, which had changed the friendly character of their relationship. It was a friendship that she was no longer interested in, as can be gathered from the second letter.

> I followed your advice and I thought
> about   what
> happened   under any
> name this can be   come about
> between us. Sir be
> sure that there would probably
> have been nothing more; you are mistaken
> and your regret is groundless.
> True friendship only exists
> between people of the same sex
> especially   of
> friends   sympathizes in
> a void   the absence of
> any feeling of this kind
> my trust   but it has been
> badly hurt   you have seen
> I was unhappy   asked
> the reason; I answered that
> I had had some disappointments; that others had been the cause of them.

I thought you would have taken it
like any person before
whom a word is uttered
for these [???]
The calmness of my ideas leaves me
the freedom to judge without much
reflection the people I see
usually; it is this that allows
me to rarely regret being
mistaken or being influenced in my view of them
I do not agree with you
fee...   more than
to   expect
[??]   I thank you
sincerely.

5   A Pointless Death

STÉPHANIE's refusal devastated GALOIS, whose fervent spirit longed to find what science had denied him in his love for her. He spent the last days of his prison sentence impatiently waiting to return to his political militancy, which was all that was left for him to make his life worth living. He had lost all hope regarding STÉPHANIE. She would never love him, and mathematics, his other great love, had also betrayed him, in a way. Though convinced that his theories were exact, and that they were valid for the future of algebra, he was aware that continuing to hope they would be understood by the Paris academic establishment was sheer madness. His only hope now was his faith in republican ideals, and he was eagerly looking forward to rejoining the Friends of the People.

He was released on 29 April. However, he did not leave the FAULTRIER establishment, not having enough money to pay for his room in rue des Bernardins, neither did he have any desire to go back to his mother, who had never visited him in Sainte-Pélagie. He could have joined AUGUSTE and MICHEL CHEVALIER in the Saint-Simonian community at Ménilmontant, where he would have been received with open arms. However, the rules imposed by the two *pères* BAZARD and ENFANTIN, the leaders of the movement, though not particularly strict, were rules, nonetheless, and GALOIS wanted to be totally independent at that time. Furthermore, leaving Paris would have meant postponing his political struggle for a republic in France.

After the arrests of July 1831, the Friends of the People had been forced to leave their headquarters in rue Grenelle-Saint-Honoré, and had not met for several months. At the beginning of May 1832, a new event had spurred them to further action. MARIE-CAROLINE, duchesse DE BERRY, had returned to France. Nobody had known that the widow of CHARLES X's son, at the time of her husband's assassination, was expecting an heir. The boy, now aged twelve, was living in exile in Prague, under the guidance of a tutor of the

Fig. 25: Portrait of EVARISTE GALOIS done from memory by his brother ALFRED in 1848

highest calibre, the mathematician CAUCHY, who, by taking on this hardly satisfying task, from the intellectual point of view, was thus able to demonstrate his devotion to the Bourbon dynasty.

The legitimists, who saw LOUIS-PHILIPPE as a usurper, placed all their hopes in this boy. How was the Duchess's return to be interpreted? Did it mean that the legitimists were ready for battle? If this were the case, the republicans thought that LOUIS-PHILIPPE could be put in a difficult position, and a new revolution planned. There was no time to lose. A meeting was planned for all the members in one of their houses, at 18 rue de l'Hôpital-Saint-Louis, on 7 May. GALOIS was informed, and was warmly welcomed back to the Society, since he was well-known for his ability to spur the more luke warm spirits into action.

The need for an armed uprising was immediately accepted. All that was missing was a pretext to provoke the fury of the crowds, and a date. One idea, that was not at first taken very seriously, was that a corpse to be revenged would be very useful. A hero was needed, in whose name the people of Paris would fight, a name to shout, while firing on LOUIS-PHILIPPE's police, a name on the lips of the dying. The discussion soon heated up, and, as his companions expressed their views, GALOIS, who had not spoken up to that point, became more and more excited.

He soon asked to speak, in such an authoritative tone, that all the others immediately fell silent. He explained, movingly, that his life had become pointless. All that was left for him was to offer it to the only thing he still loved: France. The corpse they needed would be his.

All those present began to protest. GALOIS was too young to die, and, in any case, would be much more useful for the republican cause alive than dead. He would not listen. However, several weeks would have to pass, so that, if the King's police had been told about the meeting, his death would not be connected to the Society's activities. He would arrange a duel with his friend L. D., but only his opponent's pistol would be loaded. He would even leave a number of letters making the duel seem plausible. Nobody, with the exception of those present, should know about his sacrifice. The Friends of the People would only have the task of spreading the news that the duel was actually a police ambush.

The members of the Society were not unanimous over GALOIS' plan, but he insisted so much that the meeting came to a close, with the agreement that they would meet again in due course, to organize the funeral, which would provide the opportunity of inciting the people to revolt. GALOIS' funeral would have to be well organized, and nothing left to chance. The sacrifice of a life was a very high price, and the plan must not fail.

Over the next few days, GALOIS' state of mind was a mixture of excitement, expectation and fear. He would have liked to share his secret with his faithful friend AUGUSTE, but he knew that the latter would oppose his plan, and, after hurrying back to Paris, try to dissuade him. All he did was write to him, on 25 May, about his disappointment over the end of his love story with STÉPHANIE:

> How can I console myself after the source of the greatest happiness known to man came to an end in only one month? Happiness and hope are at an end, now surely consumed for the rest of my life.

His disappointments of the previous year led him to say:

> Pity never! Hatred, that's all. Anyone who does not deeply feel this hatred for the present, does not really love the future.

And, thinking of his imminent death, he added:

> I am able to cast doubts on your cruel prophecy that I shall work no more. But I must admit that it is not entirely groundless.

So as not to arouse suspicions in his friend's mind, he ended the letter with a promise:

> I shall come and visit you on 1 June. I hope that we shall see each other frequently during the first fortnight of June. I shall be leaving on 15 for the Dauphiné region.

The time of GALOIS' sacrifice was at hand. On 29 May, he came to all the necessary agreements with L. D. They would meet at dawn on the following day, near the Glacière pond in the pleasant surroundings of the Gentilly area.

The time had also come for writing the letters which would prevent anyone from suspecting the true circumstances of his death. The letters are so skillfully written that they have given rise to different versions of the events in various biographies of GALOIS. Among the most popular are that he really did fight a duel, sparked off by an argument over a woman, or that his opponent was a policeman, who wanted to remove him from the political scene. A real duel, however, would not have made him so sure of dying. This certainty comes out very clearly from the letters. The first is addressed *to all republicans*:

A Pointless Death!

Fig. 26: Portrait of General LAMARQUE as a young man

I beg my patriotic friends not to chide me for dying in any other way than for my country.

I die, the victim of a cruel coquette, and of two of her victims. My life fades away amidst trivial gossip.

Why should I die for so little, for something so despicable.

Heaven is my witness that I could do nothing other than surrender to a provocation, that I tried to ward off, with all the means I had. I repent having told such foreboding truth to men so incapable of hearing it calmly. I shall take with me to the grave a spotless conscience, untainted by lies, untainted by patriotic blood.

Adieu! Life was dear to me, for the common good.

Pardon for those who killed me. They were acting in good faith.

The second letter is addressed to N. L. (NAPOLÉON LEBON?) and V. D. (VINCENT DUCHÂTELET, or VINCENT DELAUNAY?):

Dear friends,

I have been challenged to a duel by two patriots ... I cannot refuse.

> I beg your forgiveness for not having informed either of you.
> But my adversaries had ordered me ON MY HONOUR not to inform any patriots.
> Your task is very simple: to prove that I fought against my will, that is after having tried all possible compromises, and to say whether I am able to lie even on such a trivial subject as the one in question.
> Remember me, since fortune allowed me a long enough life to make my name known to the Nation.
> I die your friend.

What strikes the reader is GALOIS' insistence on certain death. He uses the words "I die," "I must die," "I shall take to the grave," "Adieu," "they killed me." During his last hours, before going to the tragic meeting that he himself had arranged, he must have felt deep regret for not having achieved the fame that his mathematics, if it had been understood, could have brought him. His mind was still teaming with half worked out ideas. He felt the need to communicate them to someone. All he had to leave was his mathematics. He wrote a last letter to AUGUSTE, who would not really have been able to understand it, but who would not refuse his friend a last favour:

> My dear friend. I have made some new discoveries in the field of mathematical analysis.

He made a brief summary of the *mémoire* he had deposited at the Academy containing the theory that is now named after him, adding some new theorems and conjectures covering seven pages. He concluded, regretfully:

> I do not have enough time and my ideas are not sufficiently well developed in this area, which is enormous.

He added the following request:

> Ask Jacobi and Gauss, in public, to give their opinions, not on the truth, but on the importance of these theorems.

As planned, on the morning of 30 May, a gun shot, fired at a distance of 25 paces, wounded GALOIS in the abdomen. The bullet perforated his intestine in various points, but did not kill him immediately. What happened then? Did a frightened L. D. run away or rush for help?

One version has a peasant, who was on his way to market, another a former officer in the royal army, finding GALOIS who had been left lying on the edge of the road, at nine-thirty in the morning and taking him to the COCHIN hospital.

ALFRED, when told, rushed to his wounded brother's bedside, but EVARISTE would not even tell him the truth, remaining faithful to the version agreed upon with the Friends of the People, that his assailant was a member of LOUIS-PHILIPPE's police. In the early hours of Thursday 31 May, Ascension Day, peritonitis set in. Since GALOIS was near death, a priest was called, but refused to speak to him.

His last words were for ALFRED: "Don't cry," he said, "I need all my courage to die at twenty."

At ten that morning the republicans had the corpse they needed.

The following day, the news of GALOIS' death appeared briefly in all the newspapers in Paris. It was only *Le Précurseur*, a constitutionalist newspaper in Lyon, that gave more details of the event:

> A deplorable duel yesterday robbed science of a young man who inspired the brightest hopes, but whose prodigious fame is only of a political nature. Young Evariste Galois ... fought a duel with an old friend, a very young man like him, and like him a member of the *Société des Amis du Peuple* ...
>
> At point blank range, each of them was given a pistol and fired.
>
> Only one of the pistols was loaded.

A few days before, DENUARD, one of the Friends of the People, had rented an apartment at 20 rue Saint-André-des-Arcs, so as to hold a new meeting. The police, who were keeping a close eye on the Society, became suspicious and had the door sealed. On 1 June, the republicans broke the seals and met to decide on the last arrangements for the revolt. The men of GISQUET, the prefect of police, raided the apartment during the meeting, arresting about thirty republicans, while the rest managed to escape. The following morning, at midday, in the cemetery of Montparnasse, roughly 3,000 people were present, ready to attack the police, as soon as the coffin had been lowered into the grave. The National Guard artillery were also on the alert.

While PLAIGNOL and PINEL, the leaders of the Friends of the People, were delivering their funeral orations, in honour of GALOIS, the word began to be passed round that General LAMARQUE had died. MAXIMILIEN LAMARQUE had been appointed a Marshall of France by NAPOLEON, on the latter's death bed. Calculations were quickly made. There would be a much larger, and more emotionally involved crowd at the old general's funeral. Why not take up the opportunity offered by this second corpse, and put the uprising off for another couple of days? The decision was swiftly taken, and the funeral came to a hasty, silent end. EVARISTE GALOIS' death had been pointless.

The *mémoire* rejected by the Academy was only published fourteen years later. It is now considered the foundation of modern algebra.

Fig. 27: Cenotaph in honour of EVARISTE GALOIS in the cemetery of Bourg-la-Reine. The GALOIS family tomb of a similar size, though in rather bad condition, can be seen in the background.

The Mathematical Work of EVARISTE GALOIS

EVARISTE GALOIS' contribution to mathematics may, at first sight, not appear to be very substantial, if we limit ourselves to merely counting the approximately 100 pages of his extant work. It is the implications they contain for further development that mark their far-reaching importance and originality. GALOIS' fragmentary and densely synthetic writings are particularly difficult to read, to the extent that some passages still await a wholly satisfactory interpretation. Only five of GALOIS' papers were published during his lifetime, the rest being published after his untimely death. Undoubtedly the best way of presenting GALOIS' work is to follow the author's own subdivision to be found in the letter to AUGUSTE CHEVALIER, written on 29 May 1832.

I did several new things concerning analysis. Some of them are about the theory of equations, others about integral functions.

Concerning the theory of equations, I have tried to find out under what circumstances equations are solvable by radicals, which gave me the opportunity of investigating thoroughly, and describing, all transformations possible on an equation, even if it is the case that it is not solvable by radicals.

With this material I shall be able to write three mémoires.

1

Probably GALOIS does not mention his first paper considering it as a mere student exercise. Actually his paper, entitled "Démonstration d'un théorème sur les fractions continues périodiques" and published in GERGONNE's *Annales de Mathématiques*, volume 19 (1828–29), even if of some interest, has nothing to do with the boldness of ideas that characterises the rest of the young French genius' production.

In the *mémoire* "Introductio in analysin infinitorum" (1748) L.

EULER had first demonstrated that if the coefficients of a second degree algebraic equation are rational numbers, then any root of the equation itself can be developed as a periodic continuous fraction with the partial numerators equal to 1 and the partial denominators positive; and vice versa. This question was considered again by J. LAGRANGE in 1770 in his *Additions au mémoire sur la résolution des équations numériques*.

Having a good knowledge of LAGRANGE's works, in his paper GALOIS proves the following theorem:

If one of the roots of an equation of any degree is an immediately periodic continuous fraction, then this equation will necessarily have another equally periodic root, that will be obtained by dividing the negative unit by the same periodic continuous fraction, written in inverse order.

In his demonstration of the theorem GALOIS limits himself to considering only four term periods "because the uniform course of calculus proves that if we admitted a greater number [of terms] then nothing would change."

Therefore, given the root

$$A = a + \cfrac{1}{b + \cfrac{1}{c + \cfrac{1}{d + \cfrac{1}{a + \cfrac{1}{b + \ldots}}}}}$$

there will be another one of the following type

$$-\frac{1}{B} = -\cfrac{1}{d + \cfrac{1}{c + \cfrac{1}{b + \cfrac{1}{a + \cfrac{1}{d + \cfrac{1}{c + \ldots}}}}}}.$$

Later on, GALOIS proves that:

$$\text{if} \quad A > 1 \quad \text{then} \quad -1 < -\frac{1}{B} < 0$$

and
$$\text{if } -1 < A < 0 \quad \text{then} \quad -\frac{1}{B} > 1 .$$

Thus:

When one of the roots of a second degree equation is an immediately periodic continuous fraction, greater than the unit, the other is necessarily between 0 and -1; and, conversely, if one of the roots is between 0 and -1, then the other will be necessarily positive and greater than the unit.

It can be proved that, conversely, if one of the roots of a second degree equation is positive and greater than the unit and the other is between 0 and -1, then it will be possible to express these roots as immediately periodic continuous fractions.

After his demonstrations GALOIS notices that

If one of the roots of a second degree equation is not only immediately periodic, but also symmetric, that is the terms of the period are equal at equal distance from the extremes, then we shall have $B = A$; so that the two roots will be A and $-1/A$; so, the equation will be
$$Ax^2 - (A^2 - 1)x - A = 0 .$$

Conversely, any second degree equation of the type
$$ax^2 - bx - a = 0$$

will have immediately periodic and symmetric roots.

The paper ends with an application of what has been stated to the numerical case $3x^2 - 16x + 18 = 0$.

2

Returning to the letter to CHEVALIER, we notice that GALOIS does not supply any information concerning his first *mémoire*, limiting himself to asserting:

The first one is written, and, in spite of what Poisson said, I shall preserve it, with the corrections I made.

He is talking about the *mémoire* submitted several times to the Académie des Sciences concerning the circumstances surrounding which we amply dealt with in previous chapters. The *mémoire* formulates what is today known as GALOIS theory, upon which almost all subsequent algebraic studies can be regarded as based. The manuscript is preserved, together with the author's other papers, as Ms. 2108 of the Library of the *Institut de France*, in Paris. The text consists of a "Discours préliminaire," dated September 1830, three large double folios measuring cm 25×38 (about 10×14 inches), bearing the date 16 January 1831 and author's signature, about ten fragments and a rough copy of the first proposition and fifth proposition.

In his "Discours préliminaire", GALOIS warns the reader of the difficulties he is going to find in reading the *mémoire*.

We can assert that there is, in pure Analysis, no obscurer subject and, maybe, no subject more isolated from all the rest. This subject made necessary the use of new denominations, of new characters. We do not doubt that this snag will irritate the reader, from the very first passages. He will hardly forgive the authors benefiting from all his trust for speaking a new language to him. Yet, finally, we have been forced to conform to the necessities of the subject; its importance deserves, perhaps, some attention.

Given an algebraic equation with any coefficients, numerical or literal, recognising whether the roots can be expressed by radicals, this is the problem to which we offer a complete solution.

If now you give me an equation chosen as you like, and about which you want to know whether it is solvable by radicals or not, I should only indicate to you the means for answering your question, without wanting to charge either me or anybody else to do it. In a few words, calculi are inaccessible.

From this point, it could seem that there is no advantage deriving from the solution we propose.

If the problem arose normally from this point of view it would be so. But, in most cases, in applications of Algebraic Analysis, we reduce to equations of which we know all the properties from the start: properties by means of which it will be always easy to answer the question by the rules we are going to expose.

Then we can proceed to examine the contents of the *mémoire* by some salient passages: they will not always be clear to the reader, the best commentary being a possible comparison with works by authors who, in order to clarify and popularise GALOIS's thought, faced the same problems in subsequent times.

First of all, GALOIS gives the definition of a reducible equation specifying that this term denotes an equation having rational divisors; that is an equation whose first member can be written as a product of polynomials having coefficients in the field generated by the coefficients of the equation. GALOIS specifies:

...one can agree to regard as rational all rational functions of a certain number of determined quantities, supposed to be known a priori.

In one sense we can say that GALOIS uses both the notions of field and extension, but neither defining them, nor denoting them with a specific name. Concerning the extension with other quantities taken from the field generated by the coefficients of an equation, he simply says:

We shall say that these quantities are ADJOINED to the equation. With these conventions, we shall call RATIONAL any quantity which can be expressed as a rational function of the coefficients of the equation and of a certain number of ADJOINED quantities arbitrarily agreed upon.

The definition of substitution in the *mémoire* is obviously the definition given by CAUCHY, but GALOIS considers, for the first time in the history of mathematics, sets of substitutions closed with respect to the product, and he gives the name group to such a set: this is also the first time that this word appears in the mathematical literature with a meaning different from that of set. Actually, GALOIS states:

When we want to group substitutions we shall make them all proceed from the same permutation. [...] if substitutions S and T are in such a group, one is certain of having substitution ST.

After giving the initial definitions, GALOIS enumerates four lemmas; the first lemma states that an equation cannot have a root in common with a "rational" equation without being a factor of the latter equation; the second one states that, given any equation, with all roots distinct from each other, it is always possible to find a function of them such that, permuting in every possible way the m roots, it has $m!$ distinct values.

After denoting the roots with the small letters $a, b, c, \ldots$, GALOIS gives the example of the function $V = Aa + Bb + Cc + \ldots$, where $A, B, C, \ldots$ are suitably chosen integral numbers. Lemma III is a property which, as GALOIS himself reminds us, had already been enunciated by ABEL in his posthumous *mémoire* on elliptic functions. In the manuscript, on the left margin, in a different handwriting (probably that of POISSON) the following observation is reproduced:

The demonstration of this lemma is insufficient; however, it is true according to n° 100 in the memoir of Lagrange, Berlin, 1771.

Under it, in GALOIS' own handwriting, appears the sentence:

We have transcribed word-for-word the demonstration that we gave of this lemma in a memoir presented in 1830. We attach as an historical document the note ... which M. Poisson felt he needed to make upon it.

It will be evaluated.

The demonstration reported by other authors after GALOIS, in their didactic expositions of the theory, actually faithfully follows LAGRANGE's remarks; for this reason, we transcribe GALOIS' own words integrally.

When function V is chosen as indicated above, it will have the property that all roots of the given equation can be expressed as rational functions of V.

Let $V = \varphi(a, b, c, d, \ldots)$,
or $V - \varphi(a, b, c, d, \ldots) = 0$.

Let us multiply together all the similar equations which one obtains by permutating in these all the letters, leaving just the first one fixed; this will give the following expression:

$$(V - \varphi(a, b, c, d, \ldots))(V - \varphi(a, b, c, d, \ldots)) \times$$
$$\times (V - \varphi(a, b, c, d, \ldots)) \ldots ,$$

which is symmetric in b, c, d, etc., ..., and which can consequently be written as a function of a. We shall therefore have an equation of the form $F(V, a) = 0$.

But I say that one can extract from this the value of a. For this it suffices to look for the common solution of this equation and the given one: for one cannot have, for example $F(V, b) = 0$ unless (this equation having a common factor with the similar equation) one of the functions $\varphi(a, \ldots)$ is equal to one of the functions $\varphi(b, \ldots)$; which is contrary to the hypothesis.

It therefore follows that a can be expressed as a rational function of V, and it is the same for the other roots.

Finally, Lemma IV states that, after writing the equation having as roots the $m!$ values of function V, and indicating with V, V',

$V'', \ldots$ the roots of an irreducible factor of this equation, if $a = f(V)$ is one of the roots of the proposed $f(x) = 0$, then so is $f(V')$.

At this point GALOIS thinks he can begin to expose his theory. The first proposition is:

Let an equation be given whose m roots are $a, b, c, \ldots$. There will always be a group of permutations of the letters $a, b, c, \ldots$ which will have the following property:
1. *each function invariant under the substitutions of this group will be known rationally;*
2. *conversely, every function of the roots which can be determined rationally will be invariant under these substitutions.*

It is in the very demonstration of this theorem that GALOIS constructs the equation group, in particular by means of Lemma III and Lemma IV.

Actually, he considers the irreducible equation of which V is a root, and indicates as

$$\varphi V, \qquad \varphi_1 V, \qquad \varphi_2 V, \qquad \ldots, \qquad \varphi_{m-1} V$$

the roots of the proposed equation; then, he subsitutes V with the other roots of the irreducible equation, obtaining

$$\begin{array}{lllll} \varphi V', & \varphi_1 V', & \varphi_2 V', & \ldots, & \varphi_{m-1} V' \\ \varphi V, & \varphi_1 V, & \varphi_2 V, & \ldots, & \varphi_{m-1} V \\ \ldots & \ldots & \ldots & \ldots, & \ldots \\ \varphi V^{(n-1)}, & \varphi_1 V^{(n-1)}, & \varphi_2 V^{(n-1)}, & \ldots, & \varphi_{m-1} V^{(n-1)} \end{array}$$

Each row is a permutation among the roots of the proposed equation and, therefore, the passage from one row to the following one is a substitution. The group consists exactly of the substitutions permitting the passage from one row to the other.

Concluding his schema, GALOIS states:

I say this group of permutations has the stated property.

The demonstration of the theorem follows. Subsequently other authors were to repeat this demonstration in almost identical forms.

It is interesting to observe that in the whole *mémoire* there are some uncertainties in the use of the words "permutation" and "substitution." As we have just seen, GALOIS talks about a group of permutations, but in stating his theorem he also uses the expression "substitutions of the group." However, there are many erasures and additions; yet, in the margin of one folio we can still read the following sentence, even though it has been erased:

replace everywhere the word permutation with the word substitution.

GALOIS never defines a group of permutations, but he does define a group of substitutions. Actually, he does not refer to CAUCHY's symbolism $\begin{pmatrix} A_1 \\ A_2 \end{pmatrix}$, which however he must have known, and always writes only the second row of the substitution, that is A_2, the permutation which is the result of the substitution. That is, even if he has a substitution in mind, he uses the word "permutation" because he only indicates the permutation which is its result.

The second proposition, whose demonstration is extremely concise, has points implying many elements of the theory, which were clarified many years later, as we shall see in works of later authors.

It states:

If one adjoins the root r of an auxiliary irreducible equation to a given equation, then
(1) one of two things will happen: either the group of the equation will not be changed; or it will be partitioned into p groups, each belonging to the given equation respectively when one adjoins each of the roots of the auxiliary equation;
(2) these groups will have the remarkable property that one will pass from one to the other in applying the same substitution of letters to all the permutations of the first one.

The p groups, all with the same order, are conjugate one to the other.

Theorem III follows:

If one adjoins all *the roots of an auxiliary equation to an equation, the groups in theorem II will have the further property that each group contains the same substitutions.*

This means that the group of the equation is reduced to the intersection of p conjugate groups.

GALOIS simply adds to the statement of the theorem the sentence:

The proof will be found.

The fourth theorem is stated as follows:

If one adjoins the numerical *value of a certain function of its roots to an equation, the group of the equation will be reduced in such a* way as to contain no permutations other than those which leave this function invariant.

At this point the author raises the problem forming Proposition V:

When is an equation solvable by simple radicals?

There are two different versions of the continuation of Proposition V, which answers the question raised: the one related in the *mémoire* and another one, belonging to a rough copy, bearing several erasures. The latter, even if it is very similar to the official version, is a little less concise and for this reason is the one we are going to transcribe:

Let G be the group of an equation solvable by radicals. Let us look for the conditions this group will satisfy. Let us follow the sequence of the operations possible in this solution, regarding the extractions of each first degree root as distinct operations.

Let us adjoin the first radical extracted in this solution to the equation; one of the following two cases can occur: either the group of the equation will decrease because of the addition of this radical; or the group will continue to be the same, this root extraction being nothing but a simple preparation.

It will always happen that, after a certain number of root extractions, the group of the equation will decrease, otherwise the equation would be unsolvable.

So, after adjoining to the equation, if necessary, a certain number of radical quantities not decreasing its group, one will arrive at a given point where the extraction of one simple root will decrease this group.

If, once arrived at this point, there were more than one way of decreasing the equation group by means of a simple root extraction, it would be necessary, for what we shall say, to consider only one radical of a degree as less high as possible among all simple radicals such that the knowledge of each of them decreases the group of the equation.

So let p be the prime number expressing this minimum degree, in such a way that the group of the equation decreases by means of one extraction of a root of degree p.

We can always suppose, at least concerning what refers to the group of the equation, that, among the quantities we adjoined before to the equation, there is a pth root of the unit α; since this expression *is obtained by means of extractions of degree below p roots, the knowledge of it will not alter the group G of the equation from any point of view. So let V be one of the values of the radical.*

Let H be the group of the equation when this quantity is adjoined, then each function invariant with respect to the substitutions of the group H will be known.

Therefore, it is necessary to divide the group G into p similar groups H having the same substitutions.

This is the condition for the radical in question to decrease the group of the equation into a submultiple group of the order p. Now let us adjoin this radical to the equation and we shall have an equation whose group will be simply H.

For this group we shall be able to follow the same line of reasoning we have followed for the previous one, and going on this way we shall finally find this general condition.

Group G whose equation is solvable by radicals has to be divided into a prime number of similar and identical groups H.

Then this group H has to be divided into a prime number of similar and identical groups K, and so on until we obtain a certain group M which will not contain anything but a prime number of substitutions.

Conversely, if group G satisfies the previous condition, then the equation will be solvable by radicals.

GALOIS' formulation of the solvability condition by radicals is completely equivalent to the one which, using modern algebraic language, states that the solvability of the group of the equation is necessary and sufficient.

One difficulty for the reader lies in GALOIS's use of the word "group" to indicate groups as well as subgroups, cosets and quotient groups; the groups $H, K, \ldots M$ of the terms quoted above are actually cosets, and by the expression *similar and identical* GALOIS means that they have the same numbers of elements. Stating that the number of cosets into which each group is divided has to be prime is equivalent to saying that the order of each quotient obtained by dividing any group of the series by the immediately succesive one has to be prime, that is each composition factor of the series has to be prime.

The procedure of division in similar and identical groups is exemplified by the author for the case of a fourth degree general equation.

The Mathematical Work of Evariste Galois

It is easy to observe this process in the known solution of fourth degree general equations. These equations are resolved by means of a third degree equation, which itself requires the extraction of a square root. In the natural sequence of ideas, it is therefore with this square root that one must begin. But in adjoining this square root to the fourth degree equation, the group of the equation, which contains twenty-four substitutions in all, is decomposed into two which contain only twelve substitutions. If the roots are designated by a, b, c, d, here is one of these groups:

$$\begin{array}{lll} abcd, & acdb, & adcb, \\ badc, & cadb, & dacb, \\ cdab, & dbac, & bcad, \\ dcba, & bdca, & cbda. \end{array}$$

We observe that the second permutation in the central column contains a careless mistake, indeed the right permutation should not be $cabd$, but $cadb$. It is very interesting to notice that when one considers any permutation of one of the three columns as an initial permutation, only one column originates a subgroup, whereas the other two are cosets of it; while, if in each column one takes any permutation of the column itself as an initial permutation, then each column is a subgroup.

Galois goes on:

Now this group itself splits into three groups, as is indicated in Theorems II and III. Thus, after the extraction of a single third degree radical only the group

$$\begin{array}{l} abcd, \\ badc, \\ cdab, \\ dcba; \end{array}$$

remains, and this group again splits into two groups

$$\begin{array}{ll} abcd, & cdab, \\ badc, & dcba. \end{array}$$

Thus, after a simple extraction of a square root,

$$\begin{array}{l} abcd, \\ badc; \end{array}$$

Fig. 28: Scribbles by GALOIS, among which the words "Liberté, égalité, fraternité ou la mort" are legible. Fol. 179b of the manuscript

remains, which will be resolved, finally, by a simple extraction of a square root.

The following propositions, i.e., propositions VI, VII and VIII, are devoted to the application of the condition to the irreducible equations whose degree is a prime number.

First of all, GALOIS constructs for them the associated group which, designating their roots by $x_1, x_2, \ldots x_p$, we briefly describe as the one constituted by the $p(p-1)$ substitutions of the type:

$$\begin{pmatrix} x_1, & x_2, & x_3, & \ldots & x_p \\ x_{\rho(a+b)}, & x_{\rho(2a+b)}, & x_{\rho(3a+b)}, & \ldots & x_{\rho(pa+b)} \end{pmatrix},$$

where $1 \leq a \leq p-1$, $0 \leq b \leq p-1$ and $\rho(z)$ is the remainder of the division of z by p.

In this case the condition of solvability by radicals is given by the following theorem:

In order for an irreducible prime degree equation to be solvable by radicals, it is necessary and sufficient that once any two of the roots are known the others can be deduced from them rationally.

Extreme conciseness and lack of formalism characterise GALOIS' work, so much so that, according to DALMAS, one can not regard it as a real writing, but just as "... that intellectual operation, of a strong intensity, preceding final writing up."

His extraordinary intuition leads GALOIS, for example, to use the concepts of subgroup, normal subgroup and quotient group in a very casual way. Therefore in the *mémoire* there is neither suitable terminology nor definitions for them.

We have seen the use of the expression *submultiple* group. While in a fragment of a few lines, one can read:

If a group is contained in another one, the latter will be the sum of groups similar to the first one, which will be called divisor.

3

The second *mémoire* requires the reconstruction of a much more complex mosaic. It contains, as GALOIS himself states

... some somewhat peculiar applications of the theory of equations.

GALOIS regards two different questions as belonging to his second *mémoire*. The application of the condition of solvability by radicals to a class of equations he calls "primitive" and the application of the theory of algebraic equations to the modular equations of elliptic functions.

In June 1830, I contributed an analysis on the imaginary numbers of the theory of numbers, as a lemma to the theory of primitive equations solvable by radicals to the Bulletin de Férussac.

The paper designated by GALOIS "as a lemma" bears the title "Sur la théorie des nombres" and is devoted to the solution of the congruences of a variable of degree > 1, module a prime number, and to their applications to the theory of primitive equations.

Even if not quoted in the letter to CHEVALIER, there is also another paper, published in April 1830 in the same journal, which should be regarded as belonging to the second *mémoire*, both because it deals with questions related to the *mémoire* and because GALOIS himself, in his June paper, refers "to the conditions given in the *Bulletin* in the month of April." It is a very brief paper, less than two pages, entitled "Analyse d'une mémoire sur la résolution des équations," containing, in addition to the definition of primitive equation, a list of results to be demonstrated, both on primitive equations and on modular equations of elliptic functions.

A manuscript bearing the title "Des équations primitives qui sont solubles par radicaux," containing the demonstrations of the two theorems on primitive equations enunciated in the letter to CHEVALIER belongs to the second *mémoire*. The manuscript, clearly left unfinished, was published, together with that containing the three double folios of the first *mémoire*, in 1846, by J. LIOUVILLE in the *Journal de mathématiques pures et appliquées*.

Before proceeding to the analysis of the papers indicated above, let us resume reading the letter to CHEVALIER, because, before introducing primitive equations, the author gives an interesting definition. GALOIS states:

... when a group G contains another group H, the group G can be divided into groups each one of them obtained by operating a same substitution on the permutation of H, in such a way that

$$G = H + HS + HS' + \ldots$$

and it can be decomposed into groups all having the same substitutions, in such a way that

$$G = H + TH + T'H + \ldots$$

In most cases, these two decompositions do not coincide. When they do, the decomposition is called proper.

In this definition, we refer to the peculiarity of the decomposition and not to that of the subgroup, therefore the definition is not different from that of the normal subgroup that modern university teachers present to their students.

In his paper of April 1830, GALOIS begins by giving the definition of primitive equation.

The equations which, being for example of degree mn, can be decomposed into m factors of degree n, by means of only one degree m equation, are called non-primitive equations. They are Gauss's equations. Primitive equations are those not enjoying such a simplification.

GAUSS's equations come from the equations of the type $x^n - 1 = 0$, examined in section VII of the *Disquisitiones Arithmeticae* (1801). Assuming n to be an odd prime number, the equation $x^n - 1 = 0$ has the only real root $x = 1$, the other roots being all imaginary and given by the equation $X = x^{n-1} + x^{n-2} + \ldots + x + 1 = 0$. Let Ω be the set of the roots of this equation. GAUSS demonstrates that the equation $X = 0$, having as roots all the $n - 1$ primitive roots of the equation $x^n - 1 = 0$, with n prime and odd, is irreducible. Actually, this result can be generalized: the equation having as roots all the primitive roots of the binomial equation $x^n - 1 = 0$, with n any natural number, is irreducible.

It is precisely to the equations of the type $x^{n-1} + x^{n-2} + \ldots + 1 = 0$ with n prime and odd, that GALOIS refers when he speaks of "Gauss's equations" in the definition of non-primitive equations. In fact GAUSS's idea to solve the equation $X = 0$, with n odd prime number, lies in gradually decomposing the equation into a greater and greater number of equations, in such a way that the coefficients of these equations can be determined by means of equations of a degree as low as possible, until one arrives at the roots of the set Ω.

GAUSS shows that if $n - 1$ is decomposed in any integral factors, for example in prime factors $\alpha, \beta, \ldots$, then the solution of the equation $X = 0$ is reduced again to that of α equations of the degree $(n - 1)/\alpha$, whose coefficients will be determined by means of a degree α equation. Analogously, the solution of each of these equations of the degree $(n - 1)/\alpha$ is reduced again to that of β equations of the degree $(n - 1)/\alpha\beta$ thanks to a degree β equation, and so on.

So, if ν is the number of factors $\alpha, \beta, \ldots$, the search for the roots of Ω led again to the solution of ν equations of the degrees $\alpha, \beta, \ldots$ respectively.

Therefore, the $X = 0$ equations with n odd prime, analysed by GAUSS, are not only irreducible, but also satisfy the definition of non-primitive equations as given by GALOIS.

We observe that, in the definition given by GALOIS, the adjective "irreducible" is missing; anyway, the author has clear in his mind that primitive and non-primitive equations are peculiar irreducible equations, so much that, in his subsequent papers, he uses the transitivity property of the group associated with these equations.

Primitive equations can be defined today as those equations whose associated groups are primitive.

The statements of three theorems, without demonstration, follow the definition in the paper.

The first of them, very similar to Proposition VIII demonstrated in the first *mémoire*, concerns a peculiar class of primitive equations, the equations that are irreducible and prime degree.

In order for a prime degree equation to be solvable by radicals, it is necessary and sufficient that once any two of the roots are known the others can be deduced from them rationally.

Also in this case, as in the definition, the adjective "irreducible" is missing. Presumably, GALOIS perceives that, if an equation were reducible, then its group would be non-transitive and its roots would be divided into several transitivity systems, so the group of the equation would be the union of subgroups transitive on the single systems, each of which would correspond to an irreducible factor of the equation. Thus, in his study of equation solvability by radicals, GALOIS can limit himself to only taking into consideration the irreducible equations, without mentioning them explicitly.

The second theorem states that

In order for a primitive equation to be solvable by radicals, it is necessary that $m = p^v$, where p is a prime number.

The third result is very curious seen from a rigorous modern viewpoint:

Apart from the cases quoted below, in order for a primitive degree p equation to be solvable by radicals, it is necessary that once any two of the roots are known the others can be deduced from them rationally.

In fact, GALOIS finds some peculiar cases which, even if they depart, as he says "very little from the general rule," are nevertheless exceptions. The cases "eluding the previous rule" are:

1st The case of $m = p^v = 9, = 25$.
2nd The case of $m = p^v = 4$ and in general that in which, a^α being

The Mathematical Work of EVARISTE GALOIS 131

a divisor of $(p^v - 1)/p - 1$, one would have a prime and $(p^v - 1)v/a^\alpha(p - 1) = p$ (mod. a^α).

GALOIS states that, in this second case, the necessary condition will be that, once two roots are known, the others can be deduced from them not rationally, but at least by means of a number of degree p radicals, equal to the number of the divisors a^α of $(p^v - 1)/p - 1$ such that a is prime and $(p^v - 1)/a^\alpha(p - 1) \equiv p$ (mod. a^α).

GALOIS ends his paper with other applications of his theory concerning modular equations of elliptic functions:

1st Let K be the module of an elliptic function, and p a given prime number > 3; in order for the degree $p + 1$ equation which gives the different modules of the functions transformed in relation to the number p, to be solvable by radicals, one of the following two conditions is necessary: either one of the roots is rationally known, or all the roots are rational functions of each other. Of course, here this concerns only peculiar values of module K. It is evident that this does not occur in general. This rule is not applicable when $p = 5$.
2nd It should be noticed that the general modular equation of the sixth degree, corresponding to number 5, can be lowered to one of the fifth degree of which it is the reduced. On the contrary, for higher degrees, modular equations can not be lowered.

The last statement is not correct; GALOIS himself in his letter to CHEVALIER changes his mind and declares that the lowering of degree $p + 1$ to degree p is possible not only when $p = 5$ but also when $p = 7$ and $p = 11$, lowering still being impossible when $p > 11$.

It is interesting to note that, in this first paper, even if the author does not mention the possibility of associating a group to each equation, let alone the general criterion of solvability by radicals, anyway he shyly states:

All these propositions have been deduced from the theory of permutations.

In the first part of the paper dated June 1830 the author develops the theory of those numbers that were to be called "Galois imaginaries".

When one agrees to regard as null all the quantities which, in algebraic calculus, one finds multiplied for a given prime number p, and, once given this convention, one searches for the solutions of an *algebraic equation $Fx = 0$, which Gauss designates by $Fx \equiv 0$, generally one takes into consideration only the whole solutions of such problems. Led to consider the incommensurable solutions by particular investigations, I have reached some results I regard as new.*

Let such an equation or congruence, $Fx = 0$, be given, and let p be its module. Let us assume at first, for the sake of greater simplicity, that the congruence in question does not admit any incommensurable factor, that is, one cannot find three functions φx, ψx, χx such that

$$\varphi x . \psi x = Fx + p \chi x \ .$$

In this case, the congruence will not admit any whole root, and not even any smaller degree incommensurable root. Therefore, it is necessary to regard the roots of this congruence as a sort of imaginary symbols, as they do not satisfy the requirements of integers, symbols whose use, in calculus, will often be as useful as the use of the imaginary $\sqrt{-1}$ in ordinary analysis.

It is the classification of these imaginary numbers, and their reduction to the smallest possible number we shall be concerned with.

In the second part of his paper the author applies the results obtained to the theory of algebraic equations.

It is especially in the theory of permutations, where it is continuously necessary to vary the index form, that taking into consideration the imaginary roots of congruences seems to be indispensable. It gives a simple and easy method for recognizing in what case a primitive equation is solvable by radicals, as I shall try, briefly, to explain.

In his previous paper, GALOIS already stated, even if without demonstrating it, that the degree of a primitive equation solvable by radicals is the power of a prime number, so now, in the study of the solvability by radicals of primitive equations, he takes into consideration only the equations whose degree is the power of a prime number p^ν.

At first, the author assumes $\nu = 1$ and, in the light of his studies on congruences, he can state that a degree p irreducible equation is solvable by radicals if and only if, designating by x_k the p roots of the equation and giving to k the p values determined by the congruence $k^p \equiv k$ (mod. p), its associate group is formed by all the substitutions of the type:

$$\begin{pmatrix} x_k \\ x_{(ak+b)p^r} \end{pmatrix},$$

where a, b are such that $a^{p-1} \equiv 1$, $b^p \equiv b$ (mod. p) and r is an integer.

GALOIS tries to generalize this result for primitive degree p^ν equations showing that:
a primitive degree p^ν equation is solvable by radicals if and only if, designating by x_k the p^ν roots of the equation and giving to k the p^ν values determined by the congruence $k^{p^\nu} \equiv k$ (mod. p), its associate group is formed by all the substitutions of the type:

$$\begin{pmatrix} x_k \\ x_{(ak+b)p^r} \end{pmatrix},$$

where a and b are such that $a^{p^\nu-1} \equiv 1$, $b^{p^\nu} \equiv b$ (mod. p) and r is an integer.

Let us observe that this is equivalent to saying:
a primitive degree p^ν equation is solvable by radicals if and only if, designating by $x_{k,l,m\ldots}$ its p^ν roots, where $k, l, m \ldots$ are ν indices all varying from 0 to $p-1$, its associate group is formed by all the substitutions of the type:

$$\begin{pmatrix} x_{k,l,m\ldots} \\ x_{\rho(ak+bl+cm+\ldots+h),\rho(a'k+b'l+c'm+\ldots+h')} \end{pmatrix},$$

where $1 \leq a, a' \ldots, b, b' \ldots, c, c' \ldots \leq p-1$, $0 \leq h, h' \ldots \leq p-1$ and $\rho(z)$ is the remainder of the division of z by p.

GALOIS' demonstration is anything but exhaustive. However, in his letter to CHEVALIER, after affirming again the necessity of this condition, he states, concerning its sufficiency:

the condition I indicated in the Bulletin de Férussac in order for the equation to be solvable by radicals is too restricted. There are few exceptions, but they exist.

The manuscript remained unfinished, and as part of the second *mémoire*, bears no date, yet, since in it GALOIS reveals he has realized that the linearity of the substitutions of the group associated with a primitive degree p equation does not ensure the solvability of the equation by radicals, it must have been written after the paper published in June 1830.

We try to find out, in general, in what case a primitive equation is solvable by radicals. Now, we can establish a general feature based upon the degree of these equations.

This feature is:

In order for a primitive equation to be solvable by radicals, it is necessary that its degree be of the form p^v, with p prime.

And from this it will immediately follow that, when an irreducible equation whose degree could admit different prime factors has to be solved, it cannot be done except by using the method of decomposition we owe to Gauss; otherwise, the equation will be irreducible.

In order to establish the general property we have stated concerning primitive equations solvable by radicals, we can suppose that the equation to be solved is primitive, but it is no longer primitive after adjoining a simple radical.

The adjective "simple" referring to the radical means here "of a prime index," in opposition to "compound"; adjoining a simple radical means to extend the rationality field of the equation by means of a root of an equation $x^n - k = 0$ with n prime and k belonging to the rationality field of the equations.

In other words, we can suppose that, n being prime, the group of the equation is divided into n irreducible groups, conjugated and not primitive. This is why, unless the degree of the equation is prime, such a group will always appear in the sequence of the decompositions.

Denoting by G the group associated with the equation, and by H the subgroup to which G is reduced by adjoining the "simple radical," the n conjugated groups are the cosets of H in G and their number is exactly n because the index of H in G is equal to the degree of the binomial equation used to extend the rationality field; for these groups the author uses the adjectives "irreducible" and "non-primitive" until now introduced only in relation to equations.

Let N be the degree of the equation and let us suppose that, after an extraction of a root of prime degree n, the equation becomes non-primitive and is divided into Q primitive degree P equations by means of only one degree Q equation.

Decomposing the equation into Q degree P equations by means of only one degree Q equation is equivalent, in group H, to dividing the roots on which its substitutions operate into Q non-primitivity systems each containing P roots; supposing that the Q degree P

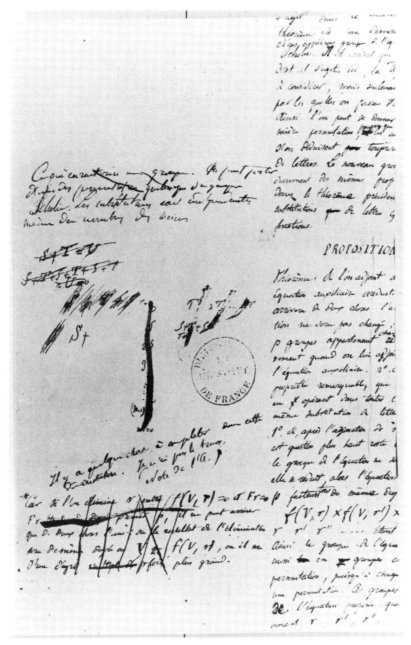

Fig. 29: The page of the *mémoire* sent to the *Académie des Sciences*, on which the words "Je n'ai pas le tem[p]s," added by GALOIS on 29 May 1832, can be read.

equations are primitive, this is equivalent to assuming that the Q non-primitivity systems consisting of P roots contain the smallest number of roots and that they cannot be divided into subsets containing less than P roots such that they are non-primitivity systems with respect to H.

If we call G the equation group, this group will have to be divided into n non-primitive conjugated groups where the letters will be arranged into systems each composed of P joined letters.

Therefore, GALOIS has clear in his mind the properties of group H associated with a non-primitive irreducible equation: the "systems each one composed of P joined letters" are the non-primitivity systems of H.

Then, the author proceeds to analyse the properties of this decomposition of the root sequence in systems, in order to see "in how many ways one will be able to do it." That is, GALOIS wants to demonstrate that, however two roots are chosen, there is one and only one decomposition of H into non-primitivity systems, each consisting of P roots, such that the two chosen roots belong to the same system.

GALOIS' demonstration is not clear at all, but, JORDAN has demonstrated later on, substantially correct even if there are too many daring passages in his reasoning.

Therefore, one will be able to distribute the N roots on which the substitutions of group H operate into Q systems, each one containing P roots.

GALOIS' conclusion that N is a power of P is anyway very rash.

The following statement gives information about what the substitutions of H look like:

So one sees, owing to the way we proceeded, that, in group H, all the substitutions will be of the type:

$$[a_{k_1,k_2,k_3,\ldots k_\mu}, a_{\varphi(k_1),\psi(k_2),\chi(k_3),\ldots \sigma(k_\mu)}]$$

because each index corresponds to a system of joined letters.

At this point, it remains to demonstrate that P is prime.

If P is not a prime number, one will reason about the group of permutations of any system of joined letters as about group G, by substituting each index with a certain number of new indices, and one will find $P = R^\alpha$, and so on; hence, finally, $N = p^\nu$, p being a prime number.

The fragment continues with the analysis of the peculiar case where the primitive equations solvable by radicals are of the degree p^2.

BOURGNE and AZRA, who were responsible for the complete edition of GALOIS's writings, published by GAUTHIERS-VILLARS in 1962, also consider another fragment GALOIS does not mention in his letter to CHEVALIER, as part of the second *mémoire*. The point of view on the question is, in this case, a little different because the author asks when a primitive group of degree pv (with p prime), a concept he has not defined, can belong to an equation solvable by radicals. What follows is, as usual, not completely clear, anyway it is based upon the following two lemmas:

Lemma I. Let G be a group of $mt \cdot n$ permutations, which can be decomposed into n groups similar to H. Let us assume that group H can be decomposed into t groups, each consisting of m permutations and similar to K.

If among all the substitutions of group G those of group H are the only ones which can transform some substitutions of group K one into another, one will have

$$n \equiv 1 \ (\mathrm{mod.}\ m) \quad \text{or} \quad tn \equiv t \ (\mathrm{mod.}\ m) \ .$$

Lemma II. If μ is a prime number and p any integer one will have

$$(x-p)(x-p^2)(x-p^3)\ldots(x-p^{\mu-1}) \equiv \frac{x^\mu - 1}{x - 1} \ \left(\mathrm{mod.}\ \frac{p^\mu - 1}{p - 1}\right) \ .$$

Concerning elliptic functions, the editors of GALOIS's work quoted above also add to the second *mémoire* a passage about the division of an elliptic function of the first class, consisting of two paragraphs respectively entitled "Résolution de l'équation algébrique de degré p^n en y supposant connue la valeur d'une fonction qui n'est invariable que par des substitutions linéaires" and "Division des transcendantes de première espèce en $m = p^n$ égales."

4

The material for the third *mémoire* can be found exclusively in the letter to CHEVALIER. GALOIS starts by taking into consideration the periods of an Abelian integral relative to any algebraic function. Subsequently he classifies the integrals into three kinds and states that if n indicates the number of the integrals of the first kind that are linearly independent, the number of the periods is $2n$. Then, he generalizes LEGENDRE's equation, where the periods of elliptic integrals appear. Obviously, links with the theory of algebraic equations are not absent, in fact GALOIS states:

The equation giving the division of periods into p equal parts is of the degree $p^{2n} - 1$. Its group has $(p^{2n} - 1)(p^{2n} - p) \ldots (p^{2n} - p^{2n-1})$ permutations.

The equation giving the division of an addition of n terms into p equal parts is of the degree p^{2n}. It is solvable by radicals.

In his letter to CHEVALIER, GALOIS never mentions other brief writings of his, two of which were published in his lifetime. The first one is a very short note, that appeared in June 1830 in the *Bulletin de Férrussac*, concerning the solution of numerical equations; the second one, published in GERGONNE's *Annales* in the same year, in December, consists of two distinct parts. In the first one, the following theorem is demonstrated:

Let Fx and fx be any two given functions; whatever x and h are one will have
$$\frac{F(x+h) - Fx}{f(x+h) - fx} = \varphi(k),$$
where φ is a determined function and k a quantity contained between x and $x + h$.

The second part contains some considerations allowing simplification of the theorems concerning the bending radii of curves in space.

Other short fragments, published posthumously, contain observations about the integration of linear equations, partial differential equations, the asymptotes of a curve, the principles of differential calculus and 2nd degree surfaces.

Bibliography[1]

1 GALOIS' Works

It was thanks to AUGUSTE CHEVALIER's tenacity that the *mémoire* which the Académie had turned down was printed. CHEVALIER made several copies of it, adding explanatory notes concerning a number of GALOIS' alterations and notes in the margin made on the eve of the mock duel.

The *mémoire* was almost certainly submitted to famous mathematicians, but only attracted the attention of JOSEPH LIOUVILLE, founding editor of the *Journal de Mathématiques pures et appliquées*, in 1843. LIOUVILLE published the *mémoire*, together with some other work by GALOIS, in his journal, three years later. As a result, CHEVALIER presented him with the complete collection of GALOIS' manuscripts.

LIOUVILLE died in 1882, leaving all his books and documents to his son-in-law CÉLESTIN DE BLIGNIÈRES. On the death of her husband, in 1905, Mme. DE BLIGNIÈRES took on the difficult task of searching for the GALOIS manuscripts among DE BLIGNIÈRES' papers. After collecting these manuscripts together, she decided to donate them to the *Institut de France*, at the same time authorizing JULES TANNERY to consult them. The result was the publication of works by GALOIS that had been overlooked by LIOUVILLE.

GALOIS' manuscripts (under the class mark Ms. 2108) are at present housed in the Library of the *Institut de France*. They include the school exercises kept throughout his life by RICHARD: after his death in 1849, they passed into the hands of CHARLES HERMITE, who left them, together with the rest of his library, to EMILE PICARD. It was PICARD who donated them to the *Institut de France*.

---

[1] Translations into several languages have been included, with the purpose of illustrating the reception of the original works in various countries.

A Texts Published During his Lifetime

April 1829. "Démonstration d'un théorème sur les fractions continues périodiques." *Annales de mathématiques pures et appliquées*, recueil périodique rédigé et publié par J. D. GERGONNE, t. XIX, n° 10, 294–301.

April 1830. "Analyse d'un mémoire sur la résolution algébrique des équations." *Bulletin des sciences mathématiques physique et chimiques*, rédigé par MM. STURM et GAULTIER DE CLAUVRY, 1re section du Bulletin universel publié par la Société pour la propagation des connaissances scientifiques et industrielles, et sous la direction de M. le Baron DE FÉRUSSAC, t. XIII, à Paris, chez Bachelier, quai des Grands-Augustins, n° 55, § 138, 271–272.

June 1830. "Note sur la résolution des équations numeriques." *Bulletin des Sciences mathématiques, physiques et chimiques*, rédigé etc., t. XIII, § 216, 413–414.

June 1830. "Sur la théorie des nombres." *Bulletin des sciences mathématiques, physiques et chimiques*, rédigé etc., t. III, § 218, 428–435.

December 1830. "Notes sur quelques points d'analyse." *Annales de mathématiques pures et appliquées*, t. XI, n° 6, 182–184.

January 1831. "Lettre sur l'enseignement des sciences." *Gazette des écoles, Journal de l'instruction publique, de l'université, des séminaires*, n° 110, 2e année.

B Posthumous Works, Collected Works

[Letter to Auguste Chevalier], 29 May 1832,
  Révue Encyclopédique, t. 55, septembre 1832, 568–576.
  The letter is preceded by a note by the Editors of the *Révue* and followed in the same number by the obituary by AUGUSTE CHEVALIER.

LIOUVILLE, JOSEPH (Ed.), "Œuvres mathématiques d'Évariste Galois." *Journal de mathématiques pures et appliquées*, t. XI, 1846, 381–444.
  The "Œuvres" contain GALOIS' published papers, the Letter to CHEVALIER and, for the first time, the *mémoire* rejected by the *Académie* and a fragment of a second *mémoire*. There is also a short note by LIOUVILLE.
  German translation by H. MASER, in *Abhandlungen über die algebraische Auflösung der Gleichungen von N. H. Abel und E. Galois*, Springer, Berlin, 1889, 87–140.

PICARD, EMILE (Ed.), *Œuvres mathématiques d'Évariste Galois*, publiées sous les auspices de la Société mathématique de France, Gauthier-Villars, Paris, 1897.
  The same content as LIOUVILLE's edition, but in book form and with a foreword by PICARD.

TANNERY, JULES (Ed.), "Manuscrits et papiers inédits de Galois." *Bulletin des sciences mathématiques*, (2), t. XXX, 1906, 226–248; 255–263.
  It contains:

1. a very detailed description (dimensions etc.) of the following manuscripts, but with no texts (presumably because they appear in LIOUVILLE's edition):
   [Letter to Auguste Chevalier],
   "Mémoire sur les conditions de résolubilité par radicaux,"
   "Des équations primitives qui sont solubles par radicaux";
2. the edition of the following unpublished texts, in some cases with a description of the documents, as before:
   A "Discours préliminaire,"
   B [Fragments; in BOURGNE et AZRA's edition they appear with the titles: "Project de publication"; "Note sur Abel"],
   C "Préface" [second part],
   D–E "Discussion sur les progrès de l'analyse pure,"
   English translation by HELEN M. KLINE, "Discussion on the Progress of Pure Analysis." *The American Mathematical Monthly*, 85, n. 7, 1978, 565–566.
   F [Fragment: "Ici comme dans toutes les science..."],
   G "Sciences, Hiérarchie, Écoles."

TANNERY, JULES (Ed.), "Manuscrits et papiers inédits de Galois." *Bulletin des sciences mathématiques*, (2), t. XXXI, 1907, 275–308.
It contains:
H–I [Various fragments on permutations and algebraic equations theory],
J "Comment la théorie des équations dépend de celle des permutations."
K–L [Fragments on equation groups],
M [A note on non primitive equations],
N "Addition au mémoire sur la résolution des équations,"
O [Untitled note: in BOURGNE et AZRA's edition the title is "Mémoire sur la division d'une fonction elliptique de première classe"],
P "Note I. Sur l'intégration des équations linéaires,"
Q "Recherche sur les surfaces du 2^d degré."

TANNERY, JULES (Ed.), *Manuscrits de Évariste Galois*. Gauthier-Villars, Paris, 1908, pp. 68.
The reprint, in book form, of the previous edition.

Préface (first part), in RENÉ TATON, "Les relations de Galois avec les mathématiciens de son temps." *Révue d'histoire des sciences*, 1, 1947, 123–127.

[Lettre sur l'enseignement des sciences], in ANDRÉ DALMAS, *Évariste Galois, révolutionnaire et géomètre*. Fasquelle, Paris, 1956.
DALMAS' biography contains, as appendix, *(Documents)*, this letter, printed for the first time, and also a reprint of "Discours préliminaire"; "Discussion sur les progrès de l'analyse pure"; Fragment: "Ici comme dans toutes les science..."; "Préface"; "Sciences Hiérarchie, Écoles."
There are also six private letters, printed for the first time.

PICARD, EMILE (Ed.), *Œuvres mathématiques d'Évariste Galois*. Gauthier-Villars, Paris, 1951, pp. 62; 56.

A reprint of PICARD's 1897 edition followed by a reprint of G. VERRIEST's paper, "Évariste Galois et la théorie des équations algébriques" (1934).

BOURGNE, ROBERT; AZRA, J. P. (Eds.), *Écrits et mémoires mathématiques d'Évariste Galois*, Édition critique intégrale de ses manuscrits et publications. Préface de JEAN DIEUDONNÉ. Gauthier-Villars, Paris, 1962, pp. XXI–541. This is the complete edition of all materials contained in Ms 2108 of the Library of the *Institut de France*, including many pages of isolated calculations. The majority of the significant papers had already been printed: the volume also contains GALOIS' school exercises, some letters and a small number of fragments (first drafts of well-known papers etc.) printed for the first time. We have for example:
["Note sur les équations aux dérivées partielles"];
["Asymptotes d'une courbe"];
["Principes du calcul différentiel"].
A second printing of the same volume (1976) contains 10 extra pages (XXXI–541 instead of XXI–541). There are 2 pages of *Errata* and 2 tables of concordances between the pages of the printed text, the manuscript pages and those of the *Appendix* of the work containing their description.

LIOUVILLE, JOSEPH (Ed.), *Œuvres mathématiques d'Évariste Galois*, publiées en 1846 dans le *Journal de Liouville*; suivis d'une étude par SOPHUS LIE, "Influence de Galois sur le développement des mathématiques." Gabay, Sceaux, 1989.

2 Biographies

DUPUY, PAUL: "La Vie d'Évariste Galois." *Annales scientifiques de l'École normale supérieure*, t. XIII, 1896, 197–266;
reprints: *Cahiers de la Quinzaine*, 5e série, 1903;
Éditions Jacques Gabay, Sceaux 1992, pp. 100.
English translation by AZALIE G. WHEELER, *Life of Evariste Galois*, 193?
Spanish translation by F. J. DUARTE, *La vida de Evaristo Galois*. Tipografia Americana, Caracas, 1947, pp. 81;
Italian translation by CARLO MOTTI, *Vita del Galois*. Tumminelli, Roma, 1945, pp. xxvi–141.

DALMAS, ANDRÉ: *Évariste Galois, révolutionnaire et géomètre*. Fasquelle, Paris, 1956, pp. 175; corrected and enlarged edition: Le Nouveau Commerce. 1982, pp. 182.
Russian translation: *Evarist Galya, revolyutsioner i matematik*. Moscow, 1960.

ASTRUC, ALEXANDRE: *Évariste Galois*. Flammarion, Paris, 1994, pp. 221.

3 Biographical Studies

(ANONYMOUS): "Obituary." *Magasin pittoresque*, 16, 1848, 227–228.

BELL, ERIC TEMPLE: *Men of Mathematics*. New York, 1937, 362–377 (Many reprints and translations).

BERTRAND, JOSEPH: "Sur 'La vie d'Évariste Galois' par Paul Dupuy." *Journal des savants*, juillet 1899, 389–400 (reprint: *Éloges Académiques, n. s.*, Paris, 1902, 329–345).

CHÂTELET, ALBERT: "Évariste Galois." *Cahiers Rationalistes*, n° 180, Juin–Juillet 1959.

CHEVALIER, AUGUSTE: "Nécrologie 'Évariste Galois'." *Revue Encyclopédique*, t. 55, sept. 1832, 744–754.

CIPOLLA, MICHELE: "Evaristo Galois. Nel primo centenario della morte." *Esercitazioni Matematiche*, 7, 1933, 3–9.

DAVIDSON, GUSTAV: "The Most Tragic Story in the Annals of Mathematics. The Life of Évariste Galois." *Scripta Mathematica*, 6, 1938, 95–100.

DEVAUX, PIERRE: *Les aventuriers de la science*. Gallimard, Paris, 1943, pp. 233 (WATT, AMPÈRE, GALOIS, EDISON); reprint Magnard, 1947.

HENRY, CHARLES: "Manuscrits de Sophie Germain." *Revue philosophique de la France et de l'Etranger*, 8, 1879, 619–641.

INFANTOZZI, CARLOS ALBERTO: "Sur la mort d'Évariste Galois." *Revue d'histoire des sciences*, 21, 1968, 157–160.

KOLLROS, LOUIS: *Évariste Galois*. Birkhäuser Verlag, Basel, 1949, pp. 24.

MALKIN, I.: "On the 150th Anniversary of the Birth Date of an Immortal in Mathematics." *Scripta Mathematica*, 26, 1963, 197–200.

ROTHMAN, TONY: "Genius and Biographers: The Fictionalization of Évariste Galois." *The American Mathematical Monthly*, 89, n° 2, 1982, 84–106.

ROTHMAN, TONY: "The Short Life of Évariste Galois." *Scientific American*, 246, n° 4, 1982, 136–149.

ROTHMAN, TONY: "Un météore des mathématiques: Évariste Galois." *Pour la Science*, n° 56, 1982, 80–90.

SARTON, GEORGE: "Évariste Galois." *The Scientific Monthly*, 13, 1921, 363–375 (reprints: *Osiris*, 3, 1937, 241–254; and G. SARTON, *The Life of Science: Essays in the History of Civilization*. Schuman, New York, 1948).

TATON, RENÉ: "Les relations scientifiques d'Évariste Galois avec les mathématiciens de son temps." *Revue d'histoire des sciences*, 1, 1947, 114–130.

TATON, RENÉ: "Sur les relations scientifiques d'Augustine Cauchy et d'Évariste Galois." *Revue d'histoire des sciences*, 24, 1971, 123–148.

TATON, RENÉ: "Évariste Galois." *Dictionary of Scientific Biography*, Vol. 5. Scribner's Sons, New York, 1972, 259–265.

TATON, RENÉ: "Évariste Galois et ses contemporains." In *Présence d'Évariste Galois (1811–1832)*. Publication de l'association des professeurs de mathématiques de l'enseignement public, n. 48, 1982, 5–39.
English translation by P. M. NEUMANN, "Évariste Galois and his Contemporaries." *Bulletin of London Mathematical Society*, 15, 1983, 107–118.

TATON, RENÉ: "Évariste Galois et ses biographes: de l'histoire aux légendes." *Sciences et techniques en perspective*, 26, 1993, 155–172.

TERQUEM, O.: "Biographie. Richard, Professeur." *Nouvelles annales de mathématiques*, 8, 1849, 448–452; *Note*.

TITS, JACQUES: "Évariste Galois. Son œuvre, sa vie, ses rapports avec l'Académie." *Institut de France*, 1982-10, pp. 10.

4 Fiction, Plays, Films

A Novels

ARNOUX, ALEXANDRE: *Algorithme*. Grasset, Paris, 1948, pp. 307.

BERLOQUIN, PIERRE: *Un souvenir d'enfance d'Évariste Galois*. Balland, Paris, 1974, pp. 60.

INFELD, LEOPOLD: *Whom the Gods Love. The Story of Évariste Galois*. Whittlesey House, McGraw Hill, New York, London, 1948, IX–323;
reprint: The National Council of Teachers of Mathematics, Reston, Va, 1978, XV–323. (Many translations in foreign languages.)

MONDOR, H.: "L'étrange rencontre de Nerval et de Galois." *Arts*, 7 juillet 1954.

B Plays and Film-scripts

ASTRUC, ALEXANDRE: "Évariste Galois mathématicien français." [scénario], *L'Avant-Scéne Cinema*, n. 82, 1968, 50–57.

MOLE', FRANCO: "Evaristo." *Samonà e Savelli*, 1965, pp. 80 (Italian one-act drama).

C Films

Évariste Galois, directed by ALEXANDRE ASTRUC, France, 1964, 20′.

Évariste Galois, directed by DANIÈLE BAUDRIER, FR3 Pic, France, 1984, 30′.

Non ho tempo, directed by ANSANO GIANNARELLI, REIAC film, Italy, 1973, 180′.

5 Historical Works Related to Galois' Life

ALBA, ANDRÉ; ISAAC, JULES; MICHAUD, JEAN; POUTHAS, CHARLES-HENRY: *Les Révolutions*. Hachette, Paris, 1960, pp. 350.

ARAGO, FRANÇOIS: *Histoire de ma jeunesse*. Christian Bourgois Editeur, Paris, 1985.

Bibliography

BÉRARD, AUGUSTE SIMON LOUIS: *Souvenirs historiques sur la révolution de 1830*. Perrotin, Paris, 1834, pp. 507.

BERGERON, LOUIS; FURET, FRANÇOIS; KOSELLECK, REINHART: *Das Zeitalter der europäischen Revolution 1780–1848*. Fischer Bücherei, Frankfurt, 1969.

BERTIER DE SAUVIGNY, GUILLAUME DE: *La Restauration*. Flammarion, Paris, 1955, pp. 625.

BJERKNES, CARL ANTON: *Niels Henrik Abel: tableau de sa vie et de son action scientifique*, traduction française par l'auteur. Gauthier-Villars, Paris, 1885, pp. 368.

BLANC, LOUIS: *L'Histoire de dix ans (1830–1840)*. Pagnerre, Paris 1841–1844, 5 vols.
Translations in English (by the author) 1844–45; in German 1844, in Italian 1844, in Spanish.

BORY, JEAN-LOUIS: *La Révolution de Juillet*. Gallimard, Paris 1972, pp. 736.

BURNAND, ROBERT: *La vie quotidienne en France en 1830*. Hachette, Paris 1943, pp. 253.

CABET, ETIENNE: *Révolution de 1830 et situation présente*. A. Mie, Paris, 1832, pp. 389.

COURSON, JEAN-LOUIS DE: *1830, la Révolution tricolore*. Julliard, Paris 1965, pp. 430.

DAYOT, ARMAND: *Journées révolutionnaires, 1830–1848*. Flammarion, Paris, 1897, pp. 112, 140.

DEBRAUX, PAUL ÉMILE: *Les barricades de 1830, scénes historiques*. A. Boulland, Paris 1830, pp. 361.

DONNAY, MAURICE: *Le lycée Louis-le-Grand*, Nouvelle Révue Française. Paris 1939, pp. 207.

DUMAS, ALEXANDRE: *Mes mémoires*. Calman-Lévy, Paris, 1863–65.
English translation by E. M. WALLER. Methuen, London 1907–1909, 6 vols.

DUPONT-FERRIER, GUSTAVE: *Du collège de Clermont au lycée Louis-le-Grand (1563–1920); la vie quotidienne d'un collège parisien pendant plus de trois cent cinquante ans*. E. de Boccard, Paris 1921–25, 3 vols.

ELLUL, JACQUES: *Etudes sur les mouvements libéraux et nationaux de 1830*. Paris 1832.

FABRE, AUGUSTE: *La Révolution de 1830 et le véritable parti républicain*. Toisnier-Desplaces, Paris, 1833, 2 vols.

Gazette des Ecoles. 1830, 1831.

Gazette des Tribunaux. 1831.

GIRARD, GEORGE: *Les Trois Glorieuses*. Firmin Didot, Paris, 1929, pp. 241.

GIRARD, LOUIS: *La Garde nationale (1814–1871)*. Plon, Paris, 1964, pp. 397.

GISQUET, HENRI JOSEPH: *Mémoires de M. Gisquet, ancien préfet de police, écrits par lui-même*. Meline et Cans, Bruxelles 1840–44, 4 vols.

HERRE, FRANZ: *Napoleon Bonaparte. Wegbereiter des Jahrhunderts.* C. Bertelsmann Verlag, München 1988.

HODDE, LUCIEN DE LA: *Histoire des sociétés secrètes et du parti républicain de 1830 à 1848.* Julien et Lanier, Paris, 1850, pp. 511.
English translation: *History of Secret Societies, and of the Republican Party of France from 1830–1848.* Lippincott, Philadelphia, 1856, pp. 479.

Journal des Débats. 1831.

La Tribune du Mouvement. 1832.

Le Constitutionel. 1830.

Le Globe. 1830, 1831.

Le Journal de Commerce. 1830.

LEMOINE, YVES; LENOEL, PIERRE: *Les Avenues de la République. Souvenirs de F.-V. Raspail sur sa vie et son siècle (1794–1878).* Hachette, Paris, 1984, pp. 379.

Le Moniteur. 1830.

Le National. 1832.

Le Temps. 1830.

LICHTHEIM, GEORGE: *The Origins of Socialism.* Frederick A. Praeger Publishers, New York, 1969, pp. 302.

LUCAS-DUBRETON, JEAN: *La Restauration et la Monarchie de Juillet.* Hachette, Paris, 1926, pp. 319.

MAZAS, ALEXANDRE: *Mémoires pour servir à l'histoire de la Révolution de 1830.* Paris 1833.

NERVAL, GERARD DE: "Mes Prisons." In *La bohème galante.* Michel Lévy, Paris, 1855, pp. 314.

PERREUX, GABRIEL: *Au temps des sociétés secrètes (1830–1835).* Hachette, Paris, 1931, pp. 398.

PINET, GASTON: *Histoire de l'Ecole Polytechnique.* Baudry, Paris, 1887, pp. 500.

PROST, A.: *Histoire de l'enseignement en France (1800–1967).* Colin, Paris, 1968, pp. 530.

RASPAIL, FRANÇOIS-VINCENT: *Réforme pénitentiaire. Lettres sur les prisons de Paris.* Tamisey et Champion, Paris, 1839, 2 vols.

REYNAUD, PAUL: *Les trois glorieuses.* Hachette, Paris, 1927, pp. 126.

RIBALLIER, LOUIS: *1830.* Nouvelle Librairie Nationale, Paris, 1911, pp. 316.

ROZET, LOUIS: *Chronique de Juillet 1830.* Barrois et Duprat, Paris, 1832, 2 vols.

THUREAU-DANGIN, PAUL: *Histoire de la monarchie de Juillet.* Plon et Nourrit, Paris 1884–92, 7 vols.

VIGIER, PHILIPPE: *La Monarchie de Juillet*, 4e éd. mise à jour. Presses Universitaires de France, Paris 1972, pp. 128.

WEILL, GEORGES JACQUES: *Histoire du parti républicain en France de 1814 à 1870.* Alcan, Paris 1900; 2e éd. rèfondue 1928, pp. 431.

6 Studies on Classical GALOIS Theory[2]

BETTI, ENRICO: "Sopra la risolubilità per radicali delle equazioni algebriche irriduttibili di grado primo." *Annali di Scienze Matematiche e Fisiche*, II, 1851, 5–19.
Also in *Opere Matematiche*, pubblicate per cura della R. Accademia dei Lincei, I, Milano 1903.

BETTI, ENRICO: "Sulla risoluzione delle equazioni algebriche." *Annali di Scienze Matematiche e Fisiche*, III, 1853, 49–115.
Also: *ibidem*.

BETTI, ENRICO: "Sopra la teorica delle sostituzioni." *Annali di Scienze Matematiche e Fisiche*, VI, 1855, 5–34.
Also: *ibidem*.

KRONECKER, LEOPOLD: "Über die algebraisch auflösbaren Gleichungen." *Monatsberichte der Königlich Preussischen Akademie der Wissenschaften zu Berlin vom Jahre* 1853, 367–374.
Also in *Werke*. Leipzig, Vol. I, 1895.

DEDEKIND, RICHARD: *[Eine Vorlesung über Algebra. 1856–1857.]* In W. SCHARLAU, *Richard Dedekind 1831–1981. Eine Würdigung zu seinem Geburtstag*. F. Vieweg und Sohn, Braunschweig, 1981, 59–100.
(Italian translation: *Lezioni sulla teoria di Galois*, a cura e con introduzione di LAURA TOTI RIGATELLI. Sansoni, Firenze, 1990, pp. 86.)

SERRET, JOSEPH ALFRED: *Cours d'algèbre supérieure*. Gauthier-Villars, Paris, 1866; II, Sec. V, Chapitre 5: "La résolution algébrique des équations. Recherches de Galois," 637–677.

HERMITE, CHARLES: *[Lettre á M. Serret.]* In J. A. SERRET, *Cours d'algèbre supérieure*. Gauthier-Villars, Paris, 1866; II, Sec. V, Chapitre 5: "La résolution algébrique des équations. Recherches de C. Hermite," 677–684.

JORDAN, CAMILLE: "Mémoire sur les groupes des équations résolubles par radicaux." *Comptes Rendues de l'Académie des Sciences*, LVIII, 1864, 963–966.
Also in *Œuvres*. Gauthier-Villars, Paris, 1961–1964, Vol. 1.

JORDAN, CAMILLE: "Commentaire sur la mémoire de Galois." *Comptes Rendues de l'Académie des Sciences*, LX, 1865, 770–774.
Also: *ibidem*.

JORDAN, CAMILLE: "Mémoire sur la résolution algébrique des équations." *Comptes Rendues de l'Académie des Sciences*, LXIV, 1867, 269–272, 586–590, 1179–1183.
Also: *ibidem*.

JORDAN, CAMILLE: "Lettre à M. Liouville sur la résolution algébrique des

---

[2] By "classical GALOIS theory" we mean the development of GALOIS ideas on resolubility of algebraic equations up to the modern abstract point of view, i.e., from BETTI (1851) to ARTIN (1930). The list is in chronological order.

équations." *Journal de mathématiques pures et appliquées*, (2), XII, 1867, 105–108.
Also: *ibidem*.

JORDAN, CAMILLE: "Mémoire sur la résolution algébrique des équations." *Journal de mathématiques pures et appliquées*, (2), XII, 1867, 109–157.
Also: *ibidem*.

JORDAN, CAMILLE: "Commentaire sur Galois." *Mathematische Annalen*, I, 1869, 142–160.
Also: *ibidem*.

JORDAN, CAMILLE: "Théorème sur les équations algébriques." *Comptes Rendues de l'Académie des Sciences*, LXVIII, 1869, 257–258.
Also: *ibidem*.

JORDAN, CAMILLE: "Théorèmes sur les équations algébriques." *Journal de mathématiques pures et appliquées*, (2), XIV, 1869, 139–146.
Also: *ibidem*.

JORDAN, CAMILLE: "Traité des substitutions et des équations algébriques." Gauthier-Villars, Paris, 1870, pp. 663.
Reprint: Gabay, Sceaux, 1989, pp. 667.

KÖNIG, JULIUS: "Beiträge zur Theorie der algebraischen Gleichungen." *Mathematische Annalen*, XXI, 1883, 424–433.

PETERSEN, JULIUS: *Theorie der algebraischen Gleichungen*. Host, Kopenhagen, 1878, pp. 335.
Italian translation by G. ROZZOLINO and G. SFORZA, *Teoria delle equazioni algebriche*, 2 vols. Libreria Pellerano, Napoli, 1º vol. 1891, pp. 164; 2º vol. 1892, pp. 188.

KRONECKER, LEOPOLD: "Einige Entwicklungen aus der Theorie der algebraischen Gleichungen." *Monatsberichte der Königlich Preussischen Akademie der Wissenschaften zu Berlin vom Jahre* 1879, 205–229.

KRONECKER, LEOPOLD: "Grundzüge der arithmetischen Theorie der algebraischen Grössen." *Crelle's Journal für die reine und angewandte Mathematik*, 91, 1881, 1–122.

BACHMANN, PAUL: "Ueber Galois' Theorie der algebraischen Gleichungen." *Mathematische Annalen*, XVIII, 1881, 449–468.

NETTO, EUGEN: *Substitutionentheorie und ihre Anwendungen auf die Algebra*. Teubner, Leipzig, 1882, pp. 290.
English translation by F. N. COLE, *Theory of Substitutions and Its Application to Algebra*. G. Wahr, Ann Arbor, 1892.
Italian translation by G. BATTAGLINI, *Teoria delle sostituzioni e sua applicazione all'algebra*. Loescher, Torino, 1885, pp. XII, 290.

GIUDICE, FRANCESCO: "Sulle equazioni irriducibili di grado primo risolubili per radicali." *Rendiconti del Circolo Matematico di Palermo*, (1), 1, 1887, 227–229.

GARCIA DE GALDEANO, ZOEL: *Crítica y síntesis del Algebra*. Imprenta y Librería de J. Peláez, Toledo, 1888, pp. 126.

DOLBNJA, IVAN P.: "Sur le critère de Galois concernant la résolubilité des équations algébriques par radicaux." *Nouvelles annales de mathématiques*, (3), 7, 1888, 467–485.

HÖLDER, OTTO: "Zurückführung einer beliebigen algebraischen Gleichung auf eine Kette von Gleichungen." *Mathematische Annalen*, XXXIV, 1889, 25–56.

BOLZA, OSKAR: "On the Theory of Substitution-Groups and its Application to Algebraic Equations." *American Journal of Mathematics*, XIII, 1891, 59–144.

WEBER, HEINRICH: "Die allgemeinen Grundlagen der Galois'schen Gleichungstheorie." *Mathematische Annalen*, XLIII, 1893, 521–549.

VOGT, HENRI: *Leçons sur la résolution algébrique des équations*. Librairie Nony, Paris, 1895, pp. 201.

BOREL, ÉMILE; DRACH, JULES: *Introduction à l'étude de la théorie des nombres et de l'algèbre supérieure*. Nony, Paris, 1895, Chapitres 5, 6, 7, 262–335.

PICARD, ÉMILE: *Traité d'Analyse*. Gauthier-Villars, Paris, Vol. 3, 1895, Chapitre 16, 454–523.

WEBER, HEINRICH: *Lehrbuch der Algebra*. Vieweg, Braunschweig, 1895, vol. 1, Drittes Buch, 491–698.
Reprint: Chelsea, New York, 1961.
French translation of the first volume by J. GRIESS, *Traité d'algèbre supérieure*. Gauthier-Villars, Paris, 1898, pp. 727.

ECHEGARAY, JOSÉ: *Resolución de ecuaciones y Teoría de Galois*. Imp. Fund. y Fab. de Tintas de los Hijos de J. A. Garcia, Madrid, 1897, pp. 514.

HÖLDER, OTTO: "Galois'sche Theorie mit Anwendungen." *Encyklopädie der Mathematischen Wissenschaften*, I, 1, Arithmetik und Algebra. Teubner, Leipzig, 1898–1904.

PIERPONT, JAMES: "Early History of Galois' Theory of Equations." *Bulletin of the American Mathematical Society*, (2), IV, 1898, 332–340.

BIANCHI, LUIGI: *Lezioni sulla teoria dei gruppi di sostituzioni e delle equazioni algebriche secondo Galois*. Spoerri Editore, Pisa, 1899, pp. 283.

PIERPONT, JAMES: "On Galois' Theory of Algebraic Equations." *Annals of Mathematics*, (2), I, 1899–1900, 113–143; II, 1900–1901, 22–55.

DICKSON, LEONARD EUGENE: *Introduction to the Theory of Algebraic Equations*. Wiley, New York, 1903, pp. 104.

CAJORI, FLORIAN: *An Introduction to the Modern Theory of Equations*. Macmillan, New York, 1904, pp. 230.

MAZZONI, PACIFICO: "Ricerche sulla teoria delle equazioni algebriche secondo Galois." *Rendiconti del Circolo Matematico di Palermo*, (1) 44, (1920), 1–51.

CIPOLLA, MICHELE: *Teoria dei gruppi d'ordine finito e sue applicazioni*, I, II. Circolo Matematico di Catania, Catania, 1922, pp. 259, 187.

VERRIEST, GUSTAVE: *Évariste Galois et la théorie des équations algébriques.* Gauthier-Villars, Paris, 1934, pp. 58.

BIRKHOFF, GARRETT: "Galois and Group Theory." *Osiris*, 3, 1937, 260–268.

PROCISSI, ANGIOLO: "Gli studi di Enrico Betti sulla Teoria di Galois nella corrispondenza Betti–Libri." *Bollettino dell'Unione Matematica Italiana*, VIII, 1953, 315–328.

VAN DER WAERDEN, BARTEL LEENDERT: "Die Algebra seit Galois." *Jahresbericht der Deutschen Mathematiker-Vereinigung*, LXVIII, 1966, 155–165.

WUSSING, HANS: *Die Genesis des abstrakten Gruppenbegriffes.* VEB Deutscher Verlag der Wissenschaften, Berlin, 1969, pp. 258.
English translation by A. SHENITZER, *The Genesis of the Abstract Group Concept.* The MIT Press, Cambridge, Massachusetts and London, 1984, pp. 331.

KIERNAN, B. MELVIN: "The Development of Galois Theory from Lagrange to Artin." *Archive for History of Exact Sciences*, 8, (1/2), 1971, 40–152.

VAN DER WAERDEN, BARTEL LEENDERT: "Die Galois-Theorie von Heinrich Weber bis Emil Artin." *Archive for History of Exact Sciences*, 9, 1972, 240–248.

BRASSELET, A. M.: *Résolution des équations algébriques. Premier mémoire de Galois.* Publication de l'IREM de Lille, Juin 1979.

GARMA, SANTIAGO: "La primera exposición de la teoría de Galois en España." *Llull – Boletín de la Sociedad Española de Historia de la ciencias*, 3, 1979–80, 7–14.

DAHAN-DALMEDICO, AMY: "Résolubilité des équations par radicaux et premier mémoire d'Évariste Galois." In *Présence d'Évariste Galois (1811–1832).* Publication de l'association des professeurs de mathématiques de l'enseignement public, n. 48, 1982, 43–53.

HIRANO, YOICHI: "Note sur la diffusion de la théorie de Galois: Première classification des idées de Galois par Liouville." *Historia Scientiarum*, 27, 1984, 27–41.

EDWARDS, HAROLD M.: *Galois Theory.* Springer, New York, Berlin, Heidelberg, Tokio, 1984, pp. 152.

TIGNOL, JEAN-PIERRE: *Leçons sur la théorie des équations.* Institut de Mathématique Pures et Appliquées, Louvain, 1980.
English revised translation by the author: *Galois Theory of Algebraic Equations.* Longman, Harlow, 1988, pp. 430.

TOTI RIGATELLI, LAURA: *La mente algebrica. Storia dello sviluppo della teoria di Galois nel XIX secolo.* Bramante, Busto Arsizio, 1989, pp. 170.

MAMMONE, PASQUALE: "Sur l'apport d'Enrico Betti en théorie de Galois." *Bollettino di Storia delle Scienze Matematiche.* IX, n. 2, 1989, 143–169.

7 Studies on GALOIS' Second *Mémoire*

JORDAN, CAMILLE: "Mémoire sur le nombre des valeurs des functions." *Journal de l'École Polytechnique*, XXII, 1861, 113–194.

JORDAN, CAMILLE: "Sur la résolution algébrique des équations primitives de degré p^2, (p étant premier impair)." *Journal de Mathématiques pures et appliquées*, (2), XIII, 1868, 111–135.

VALENTINI, VALENTINA: *La seconda memoria di E. Galois*. Unpublished Degree dissertation, University of Siena, 1991.

Index of Names

In this index some of the first names are missing. This is because in the documents consulted only second names are given.

ABEL, NIELS HENRIK, 38, 40, 48F., 89, 101, 119
ADELAIDE OF SAVOY, 15
AMPÈRE, general inspector of French state education, 31
AMPÈRE, ANDRÉ-MARIE, 38
ANDRY, soldier of the National Guard, 77
ARAGO, ETIENNE, 82
ARAGO, FRANÇOIS, 61, 91
ARTOIS, comte D' (see CHARLES X), 18F., 25
AUDOUIN, witness, 87
AZRA, J. P., 137

BABEUF, FRANÇOIS, 66
BACH, student at the Ecole Normale, 74F.
BALZAC, HONORÉ DE, 22
BARBIÉ DU BOCAGE, JEAN-DENIS, 47
BARBIER, member of the Society of Friends of the People, 95
BARTHE, Minister of Justice, 95
BARY, student at the Ecole Normale, 73
BASTIDE, captain of the National Guard, 68, 95
BAUDE, prefect of police, 104
BAZARD, ARMAND, 107
BELL, ERIC TEMPLE, 49
BÉNARD, student at the Ecole Normale, 60
BERNARDIN DE SAINT-PIERRE, JACQUES-HENRI, 46
BERRY, duc DE, 19, 66
BERRY, MARIE-CAROLINE, duchesse DE, 107
BERTHELIN, student at the Ecole Polytechnique, 58

BERTHOLLET, CLAUDE-LOUIS, 46
BERTHOT, NICOLAS, 24F., 29
BÉZOUT, ETIENNE, 30
BILLARD, pharmacy student, 87
BIOT, JEAN-BAPTISTE, 31, 37
BIRET, general education inspector, 58
BISSEY, student at the Ecole Normale, 73
BLANQUI, LOUIS-AUGUSTE, 58, 66, 68, 95F.
BLIGNIÈRES, CÉLESTIN DE, 139
BLIGNIÉRES, Mme DE, 139
BONNIAS, member of the Society of Friends of the People, 95
BORDEAUX, duc DE, 66
BOSSUT, CHARLES, 30
BOUDEL, law student, 88
BOURGNE, ROBERT, 137
BOURMONT, LOUIS-AUGUSTE-VICTOR DE GHAISNES, comte DE, 53
BRAVAIS, AUGUSTE, 41
BURNOUF, EUGÈNE, 70

CAILLOT, bookseller, 79F.
CAMUS, CHARLES-LOUIS-CONSTANT, 31
CAPPELLE, student at the Ecole Normale, 73
CARNOT, HIPPOLYTE, 50F., 53
CARNOT, LAZARE, 51
CARNOT, SADI, 50
CAUCHY, AUGUSTIN-LOUIS, 37–39, 41, 48F., 51, 66, 90, 109, 119, 122
CAVAIGNAC, GODEFROY, 53, 59, 65, 68, 77, 81, 95F.
CHANTELAUZE, JEAN-CLAUDE, 55
CHANVIN, soldier of the National Guard, 77
CHAPARRE, soldier of the National Guard, 77
CHARLES X, 25–27, 42, 53, 55, 58, 63, 66, 70, 74, 88, 95F., 107
CHARRAS, JEAN-BAPTISTE-ADOLPHE, 58, 63
CHASLES, MICHEL, 35, 49
CHATEAUBRIAND, FRANÇOIS-RENÉ, 97
CHEVALIER, AUGUSTE, 49–51, 66, 80, 84, 89F., 107, 110, 112, 115, 117, 128, 131, 133, 137–139
CHEVALIER, MICHEL, 49–51, 66, 80, 89F., 107
CHOFFER, student at the Ecole Normale, 73
CHOIGNEAU, member of the Society of Friends of the People, 95

Index of Names 155

Collet, student at the Ecole Normale, 73
Comte, Auguste-François-Marie, 51
Constant, Benjamin, 88
Couet, law court official, 86
Courville de, general inspector of French state education, 46
Cousin, Victor, 47, 68, 70, 72, 79
Creton, law court official, 86
Cuper, witness, 87
Cuvier, Georges-Léopold, 90

Dabas, student at the Ecole Normale, 73
Daniel, Hippolyte, 61
Danton, soldier of the National Guard, 77
David, printer, 68
De la Hodde, Lucien, 9, 13
De Prony, Gaspard-François, 34
Delair, attorney of the Royal Law Courts, 86
Delambre, Jean-Baptiste-Joseph, 37
Delapalme, assistant public prosecutor, 95
Delaunay, Vincent, 95, 111
Demante, Adélaide-Marie, 15f.
Demante, Antoine, 46
Demante, Thomas-François, 15
Denis, waiter, 86
Denuard, member of the Society of Friends of the People, 113
Desesquelle, waiter, 86
Desforges, teacher at the Lycée Louis-le-Grand, 30, 32f.
Desmaroux, student at the Ecole Normale, 73
Dinet, examiner at the Ecole Polytechnique, 44
Dirichlet, Gustave Lejeune, 49
Drouineau, Gustave, 86
Dubourg, French General, 92
Duchâtelet, Vincent, 92–94, 111
Dufour, François-Bertrand, 92
Dumas, Alexandre, 82
Dupont, lawyer, 85, 89, 95
Duprey, student at the Ecole Normale, 73
Durandon, waiter, 86

Enfantin, Prosper, 51, 107
Euclid, 31, 101
Euler, Leonhard, 31, 102, 115

Farcy, student, 75
Faultrier, owner of private clinic, 104, 107
Foucroy, Antoine-François, 20
Fourier, Joseph, 38, 41, 49, 90
Fox, 57
Francfort, soldier of the National Guard, 77
Francœur, Louis-Benjamin, 47
Frayssinous, Denis-Luc, 46

Galois, Adélaide-Marie, 20
Galois, Alfred, 20, 44, 113
Galois, Nathalie-Théodore, 16, 20, 44, 103
Galois, Nicolas-Gabriel, 15–18, 42, 44
Galois, Théodore-Michel, 15, 44
Gardnier, soldier of the National Guard, 77
Garnier-Pagès, politician, 53
Gauss, Carl Friedrich, 37, 112, 129, 132, 134
Gérard, student at the Ecole Normale, 73
Gergonne, Joseph-Diez, 37, 49, 115, 138
Germain, Sophie, 80
Gervais, member of the Society of Friends of the People, 95
Gibbon, director of studies at the Ecole Préparatoire, 70
Girod de l'Ain, prefect of police, 104
Gisquet, Henri Joseph, 9, 13, 97, 104, 113
Gourdin, soldier of the National Guard, 77
Gourgaud, Gaspard, 59
Grivel, law student, 88
Guérard, student at the Ecole Normale, 73
Guéret, butcher, 86
Guigniault, Joseph-Daniel, 46, 60, 64, 70–75, 79
Guillard, teacher at the Lycée Louis-le-Grand, 70
Guilley, soldier of the National Guard, 77
Guinard, captain of the National Guard, 68, 77
Guinard, Céleste-Marie, 104

Hachette, Jean-Nicolas, 47, 51

HAMEL, student at the Ecole Normale, 73
HENRY, knife maker, 85
HERMITE, CHARLES, 35, 37, 139
HOLMBOE, BERNDT MICHAEL, 38
HUBERT, JEAN-LOUIS, 67F., 82, 87, 95
HUGUENIN, student at the Ecole Normale, 73

INFELD, LEOPOLD, 9, 100

JACOBI, CARL GUSTAV JACOB, 49, 112
JACQUINOT-GODARD, magistrate of the Royal Law Court in Paris, 95
JOUCHAULT, member of the Society of Friends of the People, 95

LA BOURDONNAIE, comte DE, 53
LABORIE, PIERRE-LAURENT, 29F.
LACROIX, SYLVESTRE-FRANÇOIS, 30–32, 34, 49, 80F., 90
LAFAYETTE, MARIE-JOSEPH, 13, 63, 65, 77, 81
LAFITTE, JACQUES, 53, 63
LAGRANGE, JOSEPH-LOUIS, 31, 34, 37, 45, 90, 116, 120
LAKANAL, JOSEPH, 45
LALAND, JOSEPH-JÉROME, 31
LAMARQUE, MAXIMILIEN, 113
LAMÉ, GABRIEL, 49
LAPLACE, PIERRE-SIMON DE, 34
LASSASSAIGNE, student at the Ecole Préparatoire, 51, 73
LAURENT, student at the Ecole Normale, 73
LAVISÉ, Mayor of Bourg-la-Reine, 16
LE VERRIER, URBAIN, 35
LEBASTARD, soldier of the National Guard, 77
LEBON, NAPOLÉON-AIMÉ, 66, 88, 111
LECLERC, JOSEPH-VICTOR, 47
LECOMTE, pharmacist, 87
LEFÉBURE DE FOURCY, LOUIS-ETIENNE, 44, 47, 51
LEGENDRE, ADRIEN-MARIE, 30–32, 34, 37–39, 48F., 138
LENIBLE, soldier of the National Guard, 77
LEROUX, PIERRE, 51
LEROY, CHARLES-ANTOINE-FRANÇOIS, 46F.
LIBRI, GUGLIELMO, 49, 80
LIOUVILLE, JOSEPH, 49, 128, 139
LOTHON, student at the Ecole Polytechnique, 58

LOUIS-PHILIPPE, 9, 13, 53, 65, 67F., 77, 82, 85–87, 89, 93, 95, 98, 109, 113
LOUIS VI, 15
LOUIS XIV, 21, 34, 57
LOUIS XV, 57
LOUIS XVI, 16
LOUIS XVIII, 16–18, 22, 26
LOUVEL, LOUIS-PIERRE, 19

MALLEVAL, FRANÇOIS-CHRISTOPHE, 22, 24
MALOT, medical student, 88
MANGIN, prefect of police, 55F.
MARIE, abbé, 30
MARIE-LOUISE OF AUSTRIA, 16
MARMONT, AUGUST-FRÉDÉRIC-LOUIS, 57, 61, 63, 91
MARRAST, ARMAND, 46, 82
MARTELET, mathematics teacher at the Ecole Polytechnique, 58
MARTIGNAC, JEAN-BAPTISTE, 42, 53
MATHÉ, law student, 88F.
MEHEMET ALI, 96
MÉRILHON, Minister of Education, 74
MIE, printer, 92
MILLER, public prosecutor, 85
MOHL, JULIUS, 70
MONGE, GASPARD, 34, 37, 46
MONIN, student at the Ecole Normale, 73

NANDIN, presiding judge, Court of Assizes, 85
NAPOLÉON-FRANÇOIS-JOSEPH-CHARLES, Duke of Reichstadt, 16
NAPOLÉON BONAPARTE, 15–18, 21F., 34, 53, 113
NAVIER, LOUIS, 41
NENS-LAFAIST, student at the Ecole Normale, 73

ORLÉANS, duc D' (see LOUIS-PHILIPPE), 53, 63, 65

PÉCHEUX D'HERBENVILLE, soldier of the National Guard, 77, 81
PÉCLET, JEAN-CLAUDE-EUGÈNE, 46F.
PENARD, soldier of the National Guard, 77
PERIER, CASIMIR, 63
PERON, law court official, 86
PETIT, representative of the officials of the court of primary jurisdiction, 86

Index of Names   159

PHILIPPE EGALITÉ, 53
PICARD, EMILE, 139
PIERROT, teacher at the Lycée Louis-le-Grand, 32
PINARD, student at the Ecole Préparatoire, 51
PINAUD, student at the Ecole Normale, 73
PINEL, CHARLES, 113
PINSONNIER, student at the Ecole Polytechnique, 58
PLAIGNOL, member of the Society of Friends of the People, 95, 113
PLANA, GIOVANNI ANTONIO AMEDEO, 49
PLANIOL, EUGÈNE, 86
POINSOT, LOUIS, 37, 49
POINTIS, soldier of the National Guard, 77
POISSON, SIMÉON-DENIS, 38, 41, 49, 80F., 90, 101F., 117, 120
POLIGNAC, JULES DE, 53, 56
POLLET, student at the Ecole Préparatoire, 51, 73
PONCELET, JEAN-VICTOR, 49
POTERIN-DUMOTEL, JEAN-LOUIS, 104
POTERIN-DUMOTEL, STÉPHANIE, 104F., 107, 110
POULLET DE LISLE, mathematician, 31
PRÉVOST, member of the Society of Friends of the People, 95

RASPAIL, FRANÇOIS-VINCENT, 66, 68, 82, 85, 87, 93, 95F., 98, 100
RÉMUSAT, CHARLES-FRANÇOIS, 56F.
RENDU, general inspector of French state education, 31
REYNAUD, FRANÇOIS-LÉONCE, 31
RICHARD, LOUIS-PAUL-EMILE, 34–37, 40F., 46, 77, 139
RILHEUX, member of the Society of Friends of the People, 95
RIVAIL, member of the Society of Friends of the People, 95
ROBERTSON, ETIENNE, 31
ROBESPIERRE, MAXIMILIEN, 82
ROUHIER, soldier of the National Guard, 77
ROUSSELLE, secretary of the Paris Academy, 46
ROUX, student at the Ecole Normale, 73
ROUX, wine waiter, 86
RUFFINI, PAOLO, 40

SAINT-MARC-GIRARDIN, teacher at the Lycée Louis-le-Grand, 30
SAINT-SIMON, comte DE, CLAUDE-HENRI DE ROUVROY, 50F.
SAMBUC, soldier of the National Guard, 77

Serret, Joseph-Alfred, 35
Souillard, witness, 87
Sturm, Jacques-Charles-François, 49

Taillefer, Louis-Gabriel, 22
Tannery, Jules, 100, 139
Thiers, Louis-Adolphe, 53, 56, 63
Thillaye, Jean-Baptiste-Antoine, 37, 47
Thomas, captain of the National Guard, 68
Thouret, member of the Society of Friends of the People, 95
Tourneaux, student at the Ecole Polytechnique, 58
Trélat, soldier of the National Guard, 77, 95

Vendeyes, student at the Ecole Normale, 73
Véron, Jean-Hippolyte (called **Vernier**), 31–33
Villèle, comte **de**, President of the Council of Ministers, 42
Vivien, prefect of police, 92, 104
Voltaire, François-Marie Arouet, 20

Wronski, Hoëne, 102

Zoilus, 101

List of Illustrations

Fig. Frontispiece: EVARISTE GALOIS at the age of fifteen 2

Fig. 1: Plaque on the house built on the site of GALOIS' birthplace 17

Fig. 2: LOUIS XVIII 19

Fig. 3: The façade of the *Lycée Louis-le-Grand* before rebuilding in 1885   21

Fig. 4: Pupils' uniforms at the *Lycée Louis-le-Grand* 1806–1906 24

Fig. 5: CHARLES X and his generals 26

Fig. 6: Page from one of GALOIS' school exercises in the school year 1828–29. Fol. 236a of the manuscript 36

Fig. 7: AUGUSTIN-LOUIS CAUCHY (1789–1857). An extremely productive mathematician, who made important contributions to infinitesimal analysis, by defining concepts allowing it to be treated with scientific rigour. 39

Fig. 8: Tomb of NICOLAS-GABRIEL GALOIS in the cemetery of Bourg-la-Reine 43

Fig. 9: Commemorative plaque of Mayor NICOLAS-GABRIEL GALOIS on the façade of Bourg-la-Reine Town Hall 45

Fig. 10: General view of the building containing the *Lycée Louis-le-Grand* and the *Ecole Normale*, seen from rue St-Jacques 47

Fig. 11: Montagne Ste Geneviève seen from the *Ecole Polytechnique*. In the background the church of St-Etienne-du-Mont and the Panthéon can be seen. 50

Fig. 12: The duc D'ORLÉANS as a young man 54

Fig. 13: *Ecole Polytechnique* students' uniform in 1830 59

Fig. 14: The astronomer FRANÇOIS ARAGO (1786–1853). He wrote a number of scientific treatises and an introduction to astronomy for the general reader. He was also active in politics. He made an important contribution to the abolition of slavery in the French colonies.   62

Fig. 15: Portrait of General LAFAYETTE as a young man 64

Fig. 16: The entry of the duc d'Orléans into Paris on 30 July 1830 ... 65

Fig. 17: Godefroy Cavaignac in the uniform of the National Guard .. 69

Fig. 18: Louis Philippe and his family in the gardens of the Château de Neuilly 78

Fig. 19: Alexandre Dumas 83

Fig. 20: Portrait of François-Vincent Raspail (1794–1878) as an old man. He can be considered a precursor of cellular theory and pathology, as well as one of the founders of cytochemistry. He made a special contribution to the spreading of basic notions of hygiene and medicine. 84

Fig. 21: Siméon-Denis Poisson (1781–1840). His research particularly covered celestial mechanics, electrostatics, magnetism and probability theory. 91

Fig. 22: View of the Pont Neuf 93

Fig. 23: Louis-Philippe "King of the French" 94

Fig. 24: François-Vincent Raspail at the time of his imprisonment in Ste-Pélagie 99

Fig. 25: Portrait of Evariste Galois done from memory by his brother Alfred in 1848 108

Fig. 26: Portrait of General Lamarque as a young man 111

Fig. 27: Cenotaph in honour of Evariste Galois in the cemetery of Bourg-la-Reine. The Galois family tomb of a similar size, though in rather bad condition, can be seen in the background. 114

Fig. 28: Scribbles by Galois, among which the words "Liberté, égalité, fraternité ou la mort" are legible. Fol. 179b of the manuscript 126

Fig. 29: The page of the *mémoire* sent to the *Académie des Sciences*, on which the words "Je n'ai pas le tem[p]s," added by Galois on 29 May 1832, can be read. 135

About the author

Laura Toti Rigatelli was awarded a doctoral degree in mathematics in 1966 by the University of Florence. Originally her research interests were concentrated on algebra, the discipline to which she dedicated numerous publications. After 1977 her attention switched to the history of mathematics, particularly the history of algebra.

As a professor of complementary mathematics since 1976, she has founded and directed the Research Centre of Medieval Mathematics at the University of Siena. She has authored and co-authored numerous books dealing with historical aspects of mathematics.

Presently, her interests are focused on the Galois theory and the algebra of the 14th and 15th centuries.

Vita Mathematica

Edited by
Emil A. Fellmann, Basel, Schweiz

The series Vita Mathematica traces the professional lives of many of the world's great mathematicians – from antiquity to the present day. These books are written broadly following the same pattern and take into consideration the research carried out on the history of science in recent decades. These books are not written primarily for historians of mathematics. Rather, they address students of mathematics, physics and technical sciences, teachers of these subjects and mathematicians and physicists who would like to study the cultural and historical context in which their special fields of expertise are grounded. These biographies will apeal to the general reader as well. The publication languages are German, English and French.

Previously published in this Series

- VM 1* **Georg Cantor,** von Walter Purkert und Hans Joachim Ilgauds (German)
  1987, ISBN 3-7643-1770-1

- VM 2* **Blaise Pascal,** von Hans Loeffel (German)
  1987, ISBN 3-7643-1840-6

- VM 3* **Heinrich Heesch**, von Hans-Günther Bigalke (German)
  1988, ISBN 3-7643-1954-2

- VM 4* **George Berkeley**, von Wolfgang Breidert (German)
  1989, ISBN 3-7643-2236-5

- VM 5 **Norbert Wiener**, von Pesi R. Masani (English)
  1990, ISBN 3-7643-2246-2

- VM 6 **André Weil**, Autobiographie (French)
  1991, ISBN 3-7643-2500-3

- VM 7 **Johannes Faulhaber**, von Ivo Schneider (German)
  1993, ISBN 3-7643-2919-X

- VM 8 **Christian Goldbach**, von A.P. Juškevič und Ju.Kh. Kopelevič (German)
  1993, ISBN 3-7643-2678-6

- VM 9 **Friedrich Wilhelm Bessel**, von K.K. Lawrinovic (German)
  1994, ISBN 3-7643-5113-6

- VM 10 **Bernhard Riemann**, von B. Laugwitz (German)
  1995, ISBN 3-7643-5189-6

In preparation
EUKLID
von Jürgen Schönbeck (German)

* out of print

Science Networks • Historical Studies

Edited by
Erwin Hiebert, Harvard University, Cambridge MA, USA
Hans Wussing, Universität Leipzig, Germany
in cooperation with an international editorial board

The publications in this series are limited to the fields of mathematics, physics, astronomy, physical chemistry, and their applications. The publication languages are English preferentially, German, and in exceptional cases also French. The series is primarily designed to publish monographs. However, special editions featuring collected letters as well as thematic groupings of smaller individual works or proceedings can be taken into consideration. Annotated sources and exceptional biographies might be accepted in rare cases. The series is aimed primarily at historians of science and libraries; it should also appeal to interested specialists, students, and diploma and doctoral candidates. In cooperation with their international editorial board, the editors hope to place a unique publication at the disposal of science historians throughout the world.

SN 10 Benoit, P., Chemla, K., Ritter, J.: Histoire de fractions, fractions d'histoire, 1992 (ISBN 7643-2693-X)

SN 11 Reich, K.: Die Entwicklung des Tensorkalküls. Vom absoluten Differentialkalkül zur Relativitätstheorie, 1992 (ISBN 3-7643-2814-2)

SN 12 Gorelik, G.E /Frenkel, V.Ya.: Matvei P. Bronstein and the Soviet Theoretical Physics in the Thirties, 1994 (ISBN 3-7643-2752-9)

SN 13 Vizgin, V.P.: Unified Field Theories in the first third of the 20th century Translated from the Russian by Julian B. Barbour, 1994 (ISBN 3-7643-2679-4)

SN 14 Klein, U.: Verbindung und Affinität, Die Grundlegung der neuzeitlichen Chemie an der Wende vom 17. zum 18. Jahrhundert, 1994 (ISBN 3-7643-5003-2)

SN 15 Sasaki, Ch. / Sugiura, M. / Dauben, J.W. (Eds): The Intersection of History and Mathematics, 1994 (ISBN 3-7643-5029-6)

SN 16 Yavetz, I.: From Obscurity to Enigma, The Work of Oliver Heaviside, 1872–1891, 1995 (ISBN 3-7643-5180-2)

SN 17 Corry, L.: Modern Algebra and the Rise of Mathematical Structures, 1996 (ISBN 3-7643- 5311-2)

SN 18 Hentschel, K.: Physics and National Socialism. An Anthology of Primary Sources, 1996 (ISBN 3-7643-5312-0)

SN 19 Ullmann, D.: Chladni und die Entwicklung der Akustik, 1750 - 1860, 1996 (ISBN 3-7643-5398-8)

Mathematics with Birkhäuser

A.N. Kolmogorov †, Moscow State University, Russia /
A.-A.P. Yushkevich †, Institute of History of Science and Technology, Moscow, Russia (Eds)

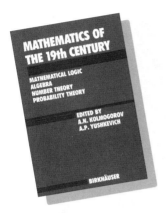

Mathematics of the 19th Century
Mathematical Logic, Algebra, Number Theory, Probability Theory

1992. 322 pages. Hardcover
ISBN 3-7643-2552-6

The history of nineteenth-century mathematics has been much less studied than that of preceding periods. The historical period covered in this book extends from the early nineteenth century up to the end of the 1930s, as neither 1801 nor 1900 are, in themselves, turning points in the history of mathematics, although each date is notable for a remarkable event: the first for the publication of Gauss' "Disquisitiones arithmeticae", the second for Hilbert's "Mathematical problems".

Beginning in the second quarter of the nineteenth century mathematics underwent a revolution as crucial and profound in its consequences for the general world outlook as the mathematical revolution in the beginning of the modern era. The main changes included a new statement of the problem of the existence of mathematical objects, particularly in the calculus, and soon thereafter the formation of non-standard structures in geometry, arithmetic and algebra.

To do justice to the vast scope of each subject, the complete work will appear in five volumes, of which the present book is the first. The multi-author work has been written by a group of specialists in the field – historians of science and mathematics - under the general direction of A.N. Kolmogorov and A.P. Yushkevich, two internationally acknowledged scientists. The primary objective of the work has been to treat the evolution of mathematics in the nineteenth century as a whole; the discussion is concentrated on the essential concepts, methods, and algorithms.

"...The book, indispensable for historians of mathematics, can be warmly recommended to every working mathematician."
EMS Newsletter No. 5, 1992

For orders originating from all over the world except USA and Canada:
Birkhäuser Verlag AG
P.O. Box 133
CH-4010 Basel / Switzerland
FAX: +41 / 61 / 205 07 92
e-mail: farnik@birkhauser.ch

For orders originating in the USA or Canada:
Birkhäuser
333 Meadowlands Parkway
USA-Secaucus, NJ 07094-2491
FAX: +1 / 800 / 777 4643
e-mail: orders@birkhauser.com

Birkhäuser
Birkhäuser Verlag AG
Basel · Boston · Berlin

MATHEMATICS WITH BIRKHÄUSER

A.N. Kolmogorov †, Moscow State University, Russia /
A.-A.P. Yushkevich †, Institute of History of Science and Technology, Moscow, Russia (Eds)

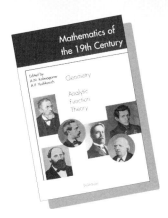

Mathematics of the 19th Century
Geometry • Analytic Function Theory

1996. 302 pages. Hardcover
ISBN 3-7643-5048-2

This book is the second volume of a study of the history of mathematics in the nineteenth century.

The first part of the book describes the development of *geometry*. The many varieties of geometry are considered and three main themes are traced: the development of a theory of invariants and forms that determine certain geometric structures such as curves or surfaces; the enlargement of conceptions of space which led to non-Euclidean geometry; and the penetration of algebraic methods into geometry in connection with algebraic geometry and the geometry of transformation groups.

The second part, on *analytic function theory*, shows how the work of mathematicians like Cauchy, Riemann and Weierstrass led to new ways of understanding functions. Drawing much of their inspiration from the study of algebraic functions and their integral, these mathematicians and others created a unified, yet comprehensive theory in which the original algebraic problems were subsumed in special areas devoted to elliptic, algebraic, Abelian and automorphic functions. The use of power series expansions made it possible to include completely general transcendental functions in the same theory and opened up the study of the very fertile subject of entire functions.

This book will be a valuable source of information for the general reader, as well as historians of science. It provides the reader with a good understanding of the overall picture of these two areas in the nineteenth century and their significance today.

For orders originating from all over the world except USA and Canada:
Birkhäuser Verlag AG
P.O. Box 133
CH-4010 Basel / Switzerland
FAX: +41 / 61 / 205 07 92
e-mail: farnik@birkhauser.ch

For orders originating in the USA or Canada:
Birkhäuser
333 Meadowlands Parkway
USA-Secaucus, NJ 07094-2491
FAX: +1 / 800 / 777 4643
e-mail: orders@birkhauser.com

Birkhäuser

Birkhäuser Verlag AG
Basel · Boston · Berlin